THE CHAOS PARADIGM: DEVELOPMENTS AND APPLICATIONS IN ENGINEERING AND SCIENCE

AIP CONFERENCE PROCEEDINGS 296

THE CHAOS PARADIGM: DEVELOPMENTS AND APPLICATIONS IN ENGINEERING AND SCIENCE

MYSTIC, CT 1993

EDITOR: RICHARD A. KATZ
NAVAL UNDERSEA
WARFARE CENTER
NEW LONDON, CT

American Institute of Physics   New York

Authorization to photocopy items for internal or personal use, beyond the free copying permitted under the 1978 U.S. Copyright Law (see statement below), is granted by the American Institute of Physics for users registered with the Copyright Clearance Center (CCC) Transactional Reporting Service, provided that the base fee of $2.00 per copy is paid directly to CCC, 27 Congress St., Salem, MA 01970. For those organizations that have been granted a photocopy license by CCC, a separate system of payment has been arranged. The fee code for users of the Transactional Reporting Service is: 0094-243X/87 $2.00.

© 1994 American Institute of Physics.

Individual readers of this volume and nonprofit libraries, acting for them, are permitted to make fair use of the material in it, such as copying an article for use in teaching or research. Permission is granted to quote from this volume in scientific work with the customary acknowledgment of the source. To reprint a figure, table, or other excerpt requires the consent of one of the original authors and notification to AIP. Republication or systematic or multiple reproduction of any material in this volume is permitted only under license from AIP. Address inquiries to Series Editor, AIP Conference Proceedings, AIP, 500 Sunnyside Boulevard, Woodbury, NY 11797-2999.

L.C. Catalog Card No. 93-074146
ISBN 1-56396-254-3
DOE CONF-9304205

Printed in the United States of America.

CONTENTS

Preface vii

MATHEMATICAL FOUNDATIONS OF CHAOS

On the Significance of Homoclinic Orbits to Chaotic Motion 3
    C. K. R. T. Jones
Applications of a Statistical Test for "Smooth" Dynamics 14
    L. W. Salvino and R. Cawley

MECHANICAL SOURCES OF CHAOS

Chaotic Sources of Noise in Machine Acoustics 27
    F. C. Moon and T. Broschart
Chaos in Gearbox Vibrations 43
    T. Frison

TURBULENCE AND CHAOS

Nonlinear Analysis of High Reynolds Number Flows Over a Buoyant
Axisymmetric Body 55
    H. D. I. Abarbanel, R. A. Katz, T. Galib, J. Cembrola, and T. W. Frison
Independent Velocity Increments and Kolmogorov's Refined Similarity
Hypotheses 97
    G. Stolovitzky and K. R. Sreenivasan
Processing of Measured Transitional and Turbulent Time Series 106
    J. Salisbury and T. A. Galib

SYNCHRONIZATION AND CONTROL OF CHAOS

Synchronizing Chaotic Circuits 127
    T. L. Carroll and L. M. Pecora
Controlling Chaos 137
    M. L. Spano and W. L. Ditto

SIGNAL MODELING AND NOISE REDUCTION STRATEGIES

Adaptive Nonlinear Dynamical Processing for Time Series Analysis 159
    J. S. Brush and J. B. Kadtke

Detection and Diagnosis of Dynamics in Time Series Data: Theory
of Noise Reduction 182
    R. Cawley, G.-H. Hsu, and L. W. Salvino
Chaotic Noise Reduction by Local-Geometric-Projection with a Reference
Time Series 193
    R. Cawley and G.-H. Hsu
Global Modeling of Chaotic Time Series with Applications to Signal
Processing 205
    J. B. Kadtke and J. S. Brush

ADVANCED APPLIED SIGNAL PROCESSING METHODS

Transient Detection Using Wavelets 233
    P. H. Carter
Dynamical Systems with Cyclostationary Orbits 246
    H. L. Hurd and C. H. Jones
Signals Associated with Nonlinear Dynamical Systems: Identification
and Monitoring 260
    J. Wright

CHAOS IN PROPAGATION MODELING AND REAL ENVIRONMENTS

Chaos in an Acoustic Propagation Model 277
    M. A. Wolfson and F. Tappert
Dimensionality, Prediction, and Determinism in the Analysis of Real
Broadband Data 289
    R. F. Wayland, Jr., D. Bromley, D. Pickett, M. E. Farrell, and A. Passamante
Author Index 299

PREFACE

These proceedings are a compilation of technical topics presented at the Second Office of Naval Research (ONR)/Naval Undersea Warfare Center (NUWC) Technical Conference on Nonlinear Dynamics and Full-Spectrum Processing. Full-spectrum processing is a phrase the Navy scientific community has adopted that entails the processing of the total available spectrum of signals of interest for use principally, but not exclusively, in passive sonar applications. However, the papers contained in these proceedings have appeal to a wide variety of applications. The conference took place in Mystic, Connecticut, on 26 and 27 April 1993 at the Seamen's Inne Conference Center, and as the conference name implies, its focus was on advances in the research of Chaos Theory combined with Applied Signal Processing.

This second conference grew out of an expanded interest in the field from the time of the first conference about two years earlier. The second conference very likely would not have taken place without the promotional efforts and inspiration provided by the ONR Defense Sciences Division (Dr. Robert J. Hansen, Director) and Physics Division (Dr. Michael Shlesinger, Director). (Dr. Hansen is currently Chief Scientist at the Applied Research Laboratory, Pennsylvania State University.) Appropriately the organizing committee members (Joan Cembrola, Tom Galib, Ted Frison, and Richard Katz) received executive approval for NUWC commendation awards for the special efforts of Dr. Hansen and Dr. Shlesinger, which were presented at the banquet session.

Another honor bestowed on our conference was the banquet presentation on the geophysical and geopolitical relevance of Nonlinear Physics to our current day experiences given by Dr. Paul Scully-Power, first Navy civilian astronaut to fly in a space shuttle mission.

Dr. Henry D. I. Abarbanel, Director of the Institute for Nonlinear Sciences, University of California, San Diego, was the lead presenter for the technical sessions. He was surrounded by an array of experts in the field (see other authors), and the technical exchanges that ensued between speaker and audience during all of the presentations are reflective of the quality of papers contained within these binders. The subject area covered by Dr. Abarbanel is contained in the October 1993 edition of *Reviews of Modern Physics* in a paper co-authored by H. Abarbanel, R. Brown, J. SIDorowich, and L. Tsimring and entitled "The Analysis of Observed Chaotic Data in Physical Systems." This book features an article by Dr. Abarbanel *et al.*, also covered in his lecture on advances in turbulence research using nonlinear processing techniques. Because of the importance of Dr. Abarbanel's results, these same experimental data were independently critiqued by Dr. John Salisbury (Analysis & Technology, Inc.) and Thomas Galib (Naval Undersea Warfare Center). Their findings are included in these proceedings and were found to be consistent with Dr. Abarbanel's results.

The conference focused on the promotion and accelerated application of Nonlinear Dynamics, Chaotic, and other advanced signal processing techniques leading to real problem solutions in engineering and the sciences. The major topics in the collection include mathematical foundations, mechanical sources of chaos, turbulence, communications, control of chaos, strategies for signal modeling and noise reduction, advanced processing methods (e.g., wavelet transform, cyclostationary, and dynamical system techniques), and chaos in channel modeling and real environments.

It requires the coordinated effort of many people behind the scenes to make a conference a successful event. The organizers wish to express sincere appreciation to Dr. William Roderick, NUWC Director of Science and Technology, NUWC Division, Newport, for providing funding support and to the conference planning/administrative support staff at Science Applications International Corporation (Douglas Brown, Fred Deltgen, and Annette Zimmermann), and to

Mr. Michael Hennelly of the American Institute of Physics for assistance in producing this publication.

The reader should find this an illuminating and provocative collection. It promotes a better understanding of Chaos Theory, which is a new mathematical paradigm for describing the nonlinear physics of nature and the applications of the theory toward solutions of practical interest in engineering and the physical sciences.

Richard A. Katz

Le Monde

...Car Dieu a fi merveilleufement eftably ces Loix, qu'encore que nous fuppofions qu'il ne crée rien de plus que ce que j'ay dit, & mefme qu'il ne mette en cecy aucun ordre ny proportion, mais qu'il en compofe vn Cahos, le plus confus & le plus embroüillé que les Poëtes puiffent décrire: elles font fuffifantes pour faire que les parties de ce Cahos fe démélent d'ellesmefmes, & fe difpofent en fi bon ordre, qu'elles auront la forme d'vn Monde tres-parfait, & dans lequel on pourra voir non feulement de la Lumiere, mais auffi toutes les autres chofes, tant generales que particulieres, qui paroiffent dans ce vray Monde.

René Descartes (1596-1650)

The World

...For God has established these Laws (of Nature) so wonderfully well that even if we suppose that he created nothing more than what I have spoken of, and even if he didn't put into it any order or proportion, but only made of it a chaos..., these laws are sufficient to make the parts of this chaos sort themselves out and dispose themselves in such good order that they will have the form of an exceedingly perfect world in which one can see not only light, but also everything else, both general and particular, which would appear in the true world.

English translation, Daniel Garber (University of Chicago)

Dr. Michael Shlesinger, Director of Physics Division of the Office of Naval Research gives opening remarks to Conference attendees.

CDR. Robert Holland, of Naval Undersea Warfare Center, Division Newport, gives welcoming address.

Dr. William I. Roderick, the Naval Undersea Warfare Center's (NUWC'S) Science and Technology Directorate addresses conference members.

Dr. Henry D. I. Abarbanel lectures on Nonlinear Analysis of NUWC Buoyant Vehicle Turbulence Experiment.

Dr. Francis C. Moon assisted by Dr. Robert Cawley demonstrates chaotic motion in a twisted metallic tape measure.

(From left to right) Francis C. Moon, Michael Shlesinger, Henry D. I. Abarbanel, Jeffrey L. Cipolla, Jeffrey S. Brush, Ted W. Frison.

Conferees mingle in discussion groups, and during coffee breaks.

MATHEMATICAL FOUNDATIONS OF CHAOS

ON THE SIGNIFICANCE OF HOMOCLINIC ORBITS TO CHAOTIC MOTION

C.K.R.T. Jones

Division of Applied Mathematics
Brown University, Providence, Rhode Island 02912

ABSTRACT

The geometry of homoclinic orbits is known to be closely related to the chaotic behavior of a system. A homoclinic orbit is the intersection between a stable and unstable manifold of the same point, or orbit. Four different manifestations of an unstable manifold, and attendant homoclinic orbits, playing a role in chaos are discussed.

INTRODUCTION

The possibility of motion in deterministic systems that is unpredictable after long periods of time was recognized by Poincaré. It remained a mathematical curiosity, albeit a deep one, until the past few decades when it was appreciated by scientists in a variety of areas that such strange forms of motion could actually occur in many physical systems. This behavior, now known as chaotic, is extremely complex and has defied any complete mathematical description.

Two main approaches have been successful in giving us at least partial explanations of the phenomena. The first is quantitative, or statistical, and concerns the calculation of quantities such as Lyapounov exponents that can distinguish chaotic from non-chaotic motion and, in some sense, quantify the degree of chaos. This is the subject of the main lectures in this conference. For a general survey on the mathematical issues involved in the statistical description of chaotic motion, I refer the reader to the monograph by Ruelle [1]. The second approach involves a geometric view for the underlying dynamical system and attempts to isolate the structures in that flow that organize the observed chaotic motion. This approach is the subject of the current paper and I hope that what is meant by "organizing the chaos" will become clear in due course. The geometric theory does not supply a discriminant to distinguish different types of chaotic motion, but it does provide some explanation as to why the chaos is present.

The basic geometric objects of study are the stable and unstable manifolds of fixed points (of maps), critical points or periodic orbits (for flows i.e., solutions of an ODE). A stable manifold consists of those points that under forward iteration, or evolution, tend to the relevant fixed point (critical point or periodic orbit). An unstable manifold is the analogous object wherein backward

4 Homoclinic Orbits to Chaotic Motion

iteration, or evolution, is used. A homoclinic orbit is an orbit that lies in *both* the stable and unstable manifolds of the same point, or periodic orbit. It is an amazing fact that a great deal can be understood about chaotic behavior from the isolating the homoclinic orbits in a given system and studying their consequences. The great beauty of this fact is that the structures involved in the homoclinic orbit, namely the stable and unstable manifolds themselves, are determined locally near the asymptotic point (or periodic orbit). The effect of the intersection of the stable and unstable manifolds, in other words the homoclinic orbit, is nevertheless felt globally. The geometry of homoclinic orbits therefore offers a way in which simple local information can be extrapolated to complicated global behavior.

In this paper, I offer four examples to underscore the point that homoclinic structures can explain many aspects of chaotic motion. No attempt is made to be exhaustive here, nor to offer a complete set of references for the subject, I merely wish to give some basic examples that are hopefully suggestive.

CHAOTIC ATTRACTORS

In this section I shall restrict attention to a smooth, invertible map F of the plane. Let p be a fixed point of F, i.e. $F(p) = p$, and assume it is of saddle-type. We denote its stable manifold W^s and its unstable manifold W^u. If there is a non-degenerate homoclinic orbit, i.e. $W^u \cap W^s \neq \phi$, then both W^u and W^s suffer strange contortions. To see this is an interesting thought-experiment; it follows from the simple fact that one intersection point implies the presence of infinitely many such intersection points. Indeed if W^u crosses W^s once, it is forced to cross it again at the image of the first point under F, and again at the image of that point and so on. The result is a tangled web of W^u and W^s that already looks strange, see Figure 1.

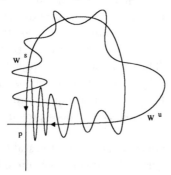

Figure 1: Homoclinic tangle

A well-known consequence of a homoclinic orbit is the presence of a set on which F restricts to a Smale horseshoe map. This set will then contain an invariant set on which the map is equivalent to the Bernoulli shift on bi-infinite sequences of 2 symbols. Since any such sequence has a corresponding point in the horseshoe invariant set and the map acts by shifting the sequence, it is not hard to see that after a large number of iterates, the behavior of the system is unpredictable. However, the horseshoe invariant set is not an attractor and cannot, in itself, account for the observed strange attractors of a dissipative system. It does, nevertheless, show that some chaotic motion is present.

This apparent shortcoming of the horseshoe invariant set as a signature of chaos in dissipative systems, and hence, by association, of homoclinic orbits, is mitigated by the fact that the full unstable manifold of some fixed point always appears to give the basic shape of the strange attractor. Moreover its contortions, as found by the presence of a homoclinic orbit, supply the "strangeness" to the attractor. Indeed, if one performs a numerical experiment iterating a point under a dissipative map of the plane that is known to exhibit chaotic behavior then, almost invariably, the long-time behavior of the orbit will track the unstable manifold. If there is a homoclinic orbit then the bending and folding of the manifold will also be evident in the numerical output. A physical example is given by the Ikeda map which models the electric field in a nonlinear, optical ring cavity, see Hammel et. al.[2] for a number of pictures that illustrate this point.

These considerations have led to the conjecture that the strange attractor in a dissipative map of the plane will always be the closure of the unstable manifold of a certain fixed point. In a recent paper of tremendous importance, this has been verified by Benedicks and Carleson[3] for the Hénon map

$$T\begin{pmatrix}x\\y\end{pmatrix} = \begin{pmatrix}1+y-ax^2\\bx\end{pmatrix} \quad (1)$$

at certain parameter values. In the following X is the closure of W^u.

Theorem (Benedicks and Carleson) For the Hénon map, there are parameter values $a, b > 0$ for which there is a positive fixed point p with the properties:

1) there is a $z \in W^u$ for which $\{T^n(z)\}_{n=0}^{\infty}$ is dense in W^u,

2) for the z given in 1), $\|DT^n(z)(0,1)\| \geq e^{cn}$ for some $c > 0$, and

3) there is an open set U for which $T^n(z)$ approaches X as $n \to +\infty$.

These properties satisfied by X are equivalent to its being a strange attractor. Property 1 implies that X cannot be decomposed into a smaller attractor. Property 2 states that there is a positive Lyapounov exponent on X and property 3 says that z is attracted to X.

This is a deep and extremely difficult result that may be very hard to generalize but it does, at least in this case, confirm the numerically corroborated

6 Homoclinic Orbits to Chaotic Motion

belief that the attractor is formed out of the unstable manifold of a certain fixed point.

CHAOTIC TRANSPORT

Since the unstable manifolds of fixed points play such a central role in the chaotic behavior of dissipative maps, it might be suspected that the stable manifold should also be significant. This is indeed the case as the stable manifolds of fixed points form the boundaries between the basins of attraction for different attractors, be they strange, fixed points or whatever. If the fixed point whose stable manifold forms such a boundary has a homoclinic orbit, then this basin boundary will become contorted. An important question arises as to how the basin boundary changes as a parameter is varied to force the onset of a homoclinic orbit at a so-called homoclinic tangency. An interesting phenomenon was discovered by Hammel and the current author[4] wherein it was shown that the basin boundary can suddenly jump as a homoclinic orbit sets in. In other words, points that were bounded away from the basin boundary at the parameter value for a homoclinic tangency can be outside that basin for all nearby parameter values at which a homoclinic orbit exists.

Attractors do not exist in Hamiltonian systems as energy and area in phase space are conserved. However, mixing of regions around the phase space does occur and is also related to the presence of homoclinic orbits. This is the subject of chaotic transport for which a beautiful theory has been developed by MacKay, Meiss and Percival[5] Wiggins[6] and others.

Consider an area preserving map of the plane. If a saddle fixed point has a homoclinic orbit then the choice of a particular, somewhat arbitrary, homoclinic point determines a simple closed curve as the union of the pieces of W^u and W^s between that homoclinic point and the fixed point, see Figure 2.

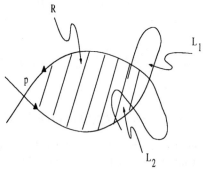

Figure 2: The region R and lobes

If this were the classical case of a 2-dimensional flow then this curve would be called a separatrix as it would separate regions of different qualitative behavior. The same idea holds in the context of maps of the plane in that we would expect different qualitative motion inside and outside the region R bounded by this curve. The question then arises as to how points might be transported between these regions. The answer lies in understanding how the homoclinic orbit forces the mapping of regions from outside to inside R. Indeed the connected regions bounded by segments of W^u and W^s, which are called lobes, are mapped to each other in an easily discernible fashion. It is not hard to see that lobe L_1 which is outside R, is mapped under F to lobe L_2, which is inside R. The homoclinic orbit has thus supplied a mechanism for traversing the boundary of R, called a turnstile by MacKay, Meiss and Percival[5]. An extensive calculus has been developed by Wiggins[6] for this lobe dynamics.

This notion of chaotic transport offers a wonderful explanation of mixing in time-dependent 2-dimensional fluid flows. Consider a 2-dimensional, incompressible fluid flow, this will be given by a velocity field $(u_1(x,y,t), u_2(x,y,t))$ and the Lagrangian trajectories are the solution paths of the ODE

$$\dot{x} = u_1(x,y,t)$$
$$\dot{y} = u_2(x,y,t). \qquad (2)$$

Suppose that (2) can be viewed as a small perturbation of a time-independent flow with a homoclinic orbit. For instance, the pendulum vector field $u_1 = y, u_2 = -\sin x$, which arises in fluids as the Kelvin cat's eye flow, see Figure 3. If the phase space is viewed with y identified modulo 2π then the separatrices become homoclinic orbits. If the small time-dependent perturbation is periodic in t then the picture for the Poincaré map is close to that of Figure 3, but the stable and unstable manifolds of the relevant fixed point may not intersect as in a map. Such an intersection can be determined by the Melnikov method. This homoclinic orbit offers effective mixing in the time-dependent flow between the region above and below the separatrix.

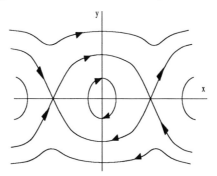

Figure 3: The pendulum phase portrait

A very interesting generalization of these transport ideas to slowly-varying perturbations has been achieved by Kaper and Wiggins[7]. The limit in this case is singular but the lobe dynamics can still be determined and they discovered an interesting phenomenon, namely that the lobe areas are actually O(1) in the limit. The lobes became progressively flatter, but also longer in the limit and the area stays finite. This was applied[7] to an eccentric journal bearing flow with inner clockwise angular velocity Ω_1 and outer counter-clockwise motion Ω_2, see Figure 4. If Ω_1 and Ω_2 are allowed to vary slowly (with respect to the time scale of the flow) then very effective mixing can take place due to the lobe areas being finite. This explains the experimental observations of Swanson and Ottino[8]. There is a striking connection between the experimental output described in that paper and the lobe dynamics in the slowly-varying theory. Indeed, the experiment shows clearly that the mixing is being carried by structures that have exactly the character of the lobes.

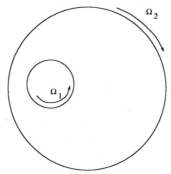

Figure 4: The eccentric journal bearing

The ideas of chaotic mixing have been applied to a 2-dimensional numerical model of the Wadden Sea by Ridderinkhof and Zimmerman[9]. The experiment is not nearly as clean as in the eccentric journal bearing but the authors show some convincing pictures which underscore the thesis that the stirring in this tidal system is by lobe dynamics.

HETEROCLINIC ATTRACTORS

In recent years there have appeared a number of studies devoted to low-dimensional approximations modelling turbulent busting near a wall in a 3-dimensional flow. The work was initiated by Aubry, Holmes, Lumley and Stone[10] and further developed by Sirovich and Zhou[11] and Sanghi and Aubry[12], among others. The idea is to reduce the problem to low-dimensions by the use of Karhunen - Loève eigenfunctions that capture the basic modes relevant to the physical problem. There is some disagreement as to how many modes are needed to capture the phenomenon and about the need for streamwise modes. There is also contention surrounding the perturbations needed to initiate the bursting. Models involving a small number of modes, however, offer an explanation of the bursting phenomenon in terms of structures of heteroclinic orbits that connect initial points. Most researchers agree that viewing this as a skeleton of the bursting mechanism offers an elegant explanation.

The simplest set of equations involve 2 roll modes and lead to the equations

$$\dot{A}_1 = \zeta_1 A_1 + \bar{A}_1 A_2 - A_1[c_{1_1}|A_1|^2 + c_{12}|A_2|^2]$$
$$\dot{A}_2 = \zeta_2 A_2 - A_1^2 - A_2[c_{2_1}|A_1|^2 + c_{22}|A_2|^2],$$
(3)

see Aubry et. al.[10]. Note that (3) has 2 easily obtained invariant subspaces: 1) A_1, A_2 both real, and 2) A_1 imaginary and A_2 real. In each of these subspaces there are heteroclinic orbits, making together a double heteroclinic loop, which is figuratively shown in Figure 5. Under certain eigenvalue conditions, namely that the unstable eigenvalue is less in absolute value than the stable one, this loop can be shown to be an attractor see Armbuster, Guckenheimer and Holmes[13] and Melbourne[14]. The swift move along the heteroclinic connection, in contrast to the slow motion near the critical points, offers an explanation of bursts in the flow. It should be noted that there is an $O(2)$ action under which the equations are invariant. The attractor is thus more complicated then the heteroclinic loops and is some non-orientable surface. However the attractor proof extends to this case, see Melbourne[14].

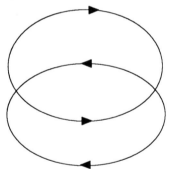

Figure 5: A heteroclinic attractor

The idea behind the proof of attraction of the heteroclinic loop is very simple, see Armbuster et. al. [13], and gives a clear example of the domination of a global structure by local behavior near a simple motion, such as a critical point. Since a finite amount of time is spent traveling between the neighborhoods of each critical point, see Figure 5, deviation from the heteroclinic loop structure will only grow by a bounded amount. However the eigenvalue condition causes a contraction onto the unstable manifold during the passage near the critical point that is exponential and far outweighs any growth during the global excursion. Indeed, suppose that the stable eigenvalue is λ_s, and the unstable eigenvalue is λ_u; the flow near one of the critical points is depicted in Figure 6.

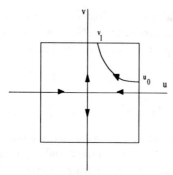

Figure 6: Flow near the critical point in a box of side δ

Suppose the trajectory enters a set of side δ at $(\pm\delta, v_0)$ and exits at $(u_1, \pm\delta)$. An easy estimate from the linear system shows that

$$u_1 \sim v_0^{\frac{\lambda_s}{\lambda_u}}.$$

If $\lambda_u < \lambda_s$ then the exponent is greater than 1 and repeated passages near the critical point will force this exponential contraction.

HOMOCLINIC MULTIPLYING

There have been many problems resolved in recent years wherein a fixed homoclinic orbit gives rise, under small perturbations, to a plethora of homoclinic orbits that lie close in the phase space. These new orbits are often doubled, or tripled etc., versions of the original orbit, that is the orbit is a sequence of excursions in the phase space, each of which is effectively the original homoclinic orbit. Evans, Fenichel and Feroe[15] found double pulses in a model nerve axon problem by a mechanism closely related to the Silnikov[16] horseshoe construction. Many studies of similar and more complicated phenomena have been performed since, including Glendenning[17], Alexander and Jones[18,19] and Mielke, Holmes and O'Reilly[20]. The motivation for the homoclinic orbits is different in each example but, in each case, there is a significant effect of homoclinic multiplying in that many new orbits appear, and some exotic structure in the phase space is strongly implied.

In his study of the validity of the quasigeostrophic approximation in atmospheric modelling, Lorenz formulated a reduced set of five equations. The goal was to understand whether, in this simple model, Rossby and gravity waves could be separated. In particular, Lorenz asked if there is a slow manifold exhibiting only Rossby wave activity. The difficulty of this issue is evident from the title of his paper[21] and that with Krishnamurthy[22]. Camassa[23] introduced a scaling and reduced, by use of an invariant, the equations to a four-dimensional system. These equations are

$$\dot{\phi} = w - bz$$
$$\dot{w} = -R^2 \sin \phi$$
$$\dot{x} = -z \tag{4}$$
$$\dot{z} = x + bR^2 \sin \phi.$$

If $b = 0$, (4) decouples into 2 separate systems, one for (ϕ, w) and one for (x, z). The (ϕ, w) - system is a pendulum equation and corresponds to Rossby wave activity. The (x, z) - system is harmonic oscillation that models the gravity waves.

Camassa analyzed pulses in the coupled system (4) for $b > 0$, but sufficiently small. He showed the presence of Rossby wave orbits asymptotic, at $\pm \infty$, to a fixed gravity wave orbit. The gravity wave orbit is a periodic orbit, and the pulse is manifested as an orbit homoclinic to that periodic orbit.

In recent work[25], a student of the author has shown a strong multiplying effect for these homoclinic orbits using the technique of the "Exchange Lemma" due to Jones and Kopell[24]. Indeed, Tin has constructed orbits with arbitrarily many Rossby wave excursions mediated by gravity wave activity. Moreover the length of time near a gravity wave can be arbitrarily long, although it must be an approximate multiple of a fixed quantity. There is certainly some unpredictability in the number of Rossby waves and in the amount of interspersed gravity wave activity. Most important, perhaps, the construction suggests that, at least in this model, Rossby and gravity waves are inextricably linked.

ACKNOWLEDGEMENTS

Research of the author has been supported by the Office of Naval Research under grant N00014-92-J-1401.

REFERENCES

1. Ruelle, D., "Chaotic Evolution and Strange Attractors", Academia Nazionale dei Lincei, Cambridge University Press Cambridge, 1989.

2. Hammel, S. M., Jones, C. and Moloney, J. V., "Global dynamical behavior of the optical field in a ring cavity", JOSA B, 2 (1985), 552–564.

3. Benedicks, M. and Carleson, L., "The dynamics of the Hénon map", Annals of Math. 133 (1991), 73–169.

4. Hammel, S. M. and Jones, C., "Jumping stable manifolds for dissipative maps of the plane", Physica 35D (1989), 87–106.

5. MacKay, R. S., Meiss, J. D., and Percival, I. C., "Transport in Hamiltanian systems, Physica 13D, 55-81.

6. Wiggins, S., "Chaotic Transport in Dynamical Systems", Springer-Verlag, New York, 1992.

7. Kaper, T. and Wiggins, S., "An analytical study of transport in Stokes flows exhibiting large-scale chaos in the eccentric journal bearing", J. Fluid Mech., to appear in 1993.

8. Swanson, P. D. and Ottino, J. M. "A comparative computational and experimental study of chaotic mixing in viscous fluids", J. Fluid Mech., 213 (1990), 227–249.

9. Ridderinkhof, H. and Zimmerman, J. T. F., "Chaotic stirring in a tidal system", Science, 258 (1992), 1107–1111.

10. Aubry, N., Holmes, P., Lumley, J. L. and Stone, E., "The dynamics of coherent structures in the wall region of turbulent boundary layer", J. Fluid Mech., 192 (1988), 115-173.

11. Sirovich, L. and Zhou, X., "Coherence and chaos in a model of turbulent boundary layer", Phys. Fluids A 4 (1992), 2855–2874.

12. Sanghi, S. and Aubry, N., "Mode interaction models for near-wall turbulence", J. Fluid Mech., 247 (1993), 455-488.

13. Armbuster, D., Guckenheimer, J. and Holmes, P., "Heteroclinic cycles and modulated travelling waves in systems with 0(2) symmetry" Physica 29D (1988), 257–282.

14. Melbourne, I., "Asymptotic stability of heteroclinic cycles in systems with symmetry", preprint (1992).

15. Evans, J. W., Fenichel, N. and Feroe, J. A., "Double impulse solutions in nerve axon equations", SIAM J. Appl. Math., 42 (1982), 219–234.

16. Silnikov, L. P., "The existence of a denumerable set of periodic motion in four-dimensional space in an extended neighborhood of a saddle-focus,", Soviet Math. Dokl. 8 (1967), 54–58.

17. Glendenning, P., "Subsidiary bifurcation near bifocal homoclinic orbits", Math. Proc. Cambridge Phil. Soc. 105 (1989), 597-605.

18. Alexander, J. C. and Jones, C., "Existence and stability of asymptotically oscillating triple pulses", Z. Angew. Math. Phys. 44 (1993), 189–200.

19. Alexander, J. C. and Jones, C., "Existence and stability of asymptotically oscillatory double pulses", to appear in J. reine angew. Math.

20. Mielke, A., Holmes, P. and O'Reilly, O., "Cascades of homoclinic orbits to chaos near a Hamiltanian saddle-center", J. Dyn. Diff. Eq. 4 (1992), 97-126.

21. Lorenz, E. N., "On the existence of a slow manifold", J. Atmos. Sci., 43 (1986), 1547–1557.

22. Lorenz, E. N. and Krishnamurthy, V., "On the nonexistence of a slow manifold", J. Atmos. Sci. 44 (1987), 2940–2950.

23. Camassa, R., "On the geometry of a slow manifold", preprint (1993).

24. Jones, C. and Kopell, N., "Tracking invariant manifolds with differential forms in singularly perturbed systems", to appear in JDE.

25. Tin, S.-K., "On homoclinic multiplying near transverse orbits bi-asymptotic to periodic orbits", preprint (1993).

APPLICATIONS OF A STATISTICAL TEST FOR "SMOOTH" DYNAMICS

Liming W. Salvino and Robert Cawley
Mathematics and Computations Branch
Naval Surface Warfare Center, Dahlgren Division, White Oak
10901 New Hampshire Avenue
Silver Spring, MD 20903-5640

ABSTRACT

We give results of a few simple applications of a statistical test for "smoothness" of embedded time series recently introduced by the authors. The method, which is applicable to both map and flow data, exploits an arbitrariness in the choice of vector field for computation of a statistic forming the basis of the test. The statistic we choose is a natural extension to the general vector field setting of Kaplan and Glass's Λ-statistic, although that specific choice is not essential to the method. Unavoidable uncertainties in Λ due to finite numerics are mitigated by the device of employing maximum and minimum values of Λ over a set of many randomly chosen vector fields. We examine properties under the test of examples chosen to illustrate the variety of effects that can occur in implementation of the test. Although we have focussed our investigations on low values of embedding trial dimension, the method seems likely to be generally reliable if appropriate data requirements are met.

INTRODUCTION

The effort to measure low dimensional dynamical behavior in irregular (aperiodic) time series is by now an old one. That it need not be an empty quest is due to the existence of chaos. This can be manifested experimentally as sustained irregularly oscillating time series which can resemble noise. Simple embedding methods [1,2] have made possible the faithful reconstruction from a single observed time series of the attractor for the underlying dynamics. So when such dynamics are there, one can sometimes hope to confirm this fact.

If an underlying dynamics is not there, however, these constructions cannot be expected to produce "embeddings" of any sort. More importantly, *unless appropriate data requirements are met*, standard fractal dimension and Lyapounov exponent measurements for noise time series can give results like those expected when dynamics are present. Hence, one can be fooled[3,4]. It is clear then that it is desirable to have a tool with which to show ahead of time that dynamics actually are present. This is one motivation for the study of the problem of distinguishing between randomness and dynamics in time series directly. Several simple approaches to the problem have appeared in the last few years[5-8]. Although the basis of our method is somewhat different theoretically, in developing it we were first motivated by the work of Kaplan and Glass[6].

We adopt the following approach: smoothness in phase space implies determinism in time series; so we have devised a statistical test whose object is to discriminate such smoothness from randomness[9]. Technically, what it detects is not smoothness, which normally refers to differentiability of some order $r > 0$, but continuity (C^0) on the reconstructed phase space. We note that the importance of "smoothness" has also been realized by Wayland, et al[8].

The time series data studied in this paper are generated for the x-coordinate for three well-known systems with standard system parameters: for the Lorenz system, $\dot{x} = 10(y-x)$, $\dot{y} = 28x - y - xz$, $\dot{z} = -\frac{8}{3}z + xy$, with sampling time $\Delta t = 0.05$ (~ 17 points per typical oscillation); for the Rössler system, $\dot{x} = -(y+z)$, $\dot{y} = x + 0.2y$, $\dot{z} = 0.2 + zx - 5.7z$, with sampling time $\Delta t = 1$ (~ 7 points per typical oscillation); and for the Ikeda map, $z' = 1.0 + 0.9z \exp\{i(0.4 - \frac{6.0}{1+|z|^2})\}$ with $z = x + iy$,. Randomized time series are constructed from the Fourier transforms (periodograms) of each of the above dynamical systems by randomizing the phases and transforming back. We refer to these random process (RP) time series as "Lorenz noise", "Ikeda noise", etc.

We report studies of our method which illustrate some of the variety of computational results that can occur in its implementation. In Section II we review the mathematical setting for our approach. In Section III we define the statistical quantities with which we work, the test itself, and applications to a couple of examples. We conclude the paper in Section IV with a summary.

MATHEMATICAL SETTING

Chaotic behavior is produced by nonlinear ordinary differential equations and maps on manifolds. As long as the right hand side of a system of ordinary differential equations is a locally Lipschitz function of position, the solutions are unique and nearby points on the phase space behave similarly under time evolution. The same things are true for a map that is a continuous functon of position. In essence, these continuity properties of a dynamical system, whether flow or map, imply unique solutions and predictable behavior, that is, they imply determinism[10].

Let a time series $v(t) : t = 1, ..., N_D$, now be the output of a differentiable dynamical system f^t on an m-dimensional manifold M; i.e., $f^t : M \to M$ and $v(t) = S(f^t p)$, where $f^0 p = p \in M$ is the initial condition and $S : M \to \mathbb{R}$ is a coordinate function on M that represents the physical observation. If the Ruelle-Takens[1,2] delay coordinate construction (DCC), which gives a vector time series in $\mathbb{R}^d$, is an embedding, the continuity properties of the dynamical system will be reproduced on the image M' of M in $\mathbb{R}^d$. The vector time series,

$$x(t) = (v(t), v(t+\Delta), ..., v(t+(d-1)\Delta)),$$

$$t = 1, ..., N = N_D - (d-1)\Delta, \qquad (1)$$

where Δ is the delay, is contained in M' and carries these same continuity properties. Let us denote the map on M' corresponding to f^1 by F and consider the following quantity,

$$\phi = \phi(x) = \Psi(x, F^b(x), ..., F^{b(R-1)}(x)), \qquad (2)$$

where Ψ is some smooth function from $\mathbb{R}^{Rd}$ into $\mathbb{R}^d$ and b is a "boost". Typically, we take $b = 1$, but for finely sampled flow data we might wish to choose $b > 1$. We assume $b = 1$ henceforth. $\phi(x)$ is a vector quantity in $\mathbb{R}^d$ varying with x. The simplest example of such a vector field is just a linear combination,

$$\phi(x) = \sum_{r=0}^{R-1} c_r F^r(x). \qquad (3)$$

F may represent a smooth map, or a smooth flow, either finely or coarsely sampled. Evidently, the c_r do not have to be constants, and may depend upon x. But for simplicity in this paper we assume the c_r to be constants.

Directional element field patterns depend on the choice of the c_r. We use Rössler and Ikeda data for illustration (Fig. 1). Note the choice $\{c_r\} = \{-1, 1\}$ produces a directional element field whose "arrows" point towards the position of the next iterate. For finely sampled flows this approximates the flow line tangent vector field.

STATISTICAL TEST

We base our statistical test on an extension of the Λ-statistic of Kaplan and Glass[6,11], viz.

$$\Lambda = \left(\sum_j n_j\right)^{-1} \sum_j n_j \frac{L_j^2 - (R_{n_j}^d)^2}{1 - (R_{n_j}^d)^2} \qquad (4)$$

where the summation extends over boxes of a grid, indexed by j, with n_j the number of data points in box-j. The quantity $R_{n_j}^d$, a constant depending only on n_j, is the expected value for the root mean square displacement in a unit step, uncorrelated random walk of n_j steps. The quantity L_j is a function of the time series data and of the choice of vector field:

$$L = L_j = n_j^{-1} \| \sum_{i=1}^{n_j} \dot{x}(x_i) \|, \qquad (5)$$

where $\dot{x} = \phi(x) \|\phi(x)\|^{-1}$ denotes a directional element (i.e. unit vector) field associated with $\phi(x)$ [12].

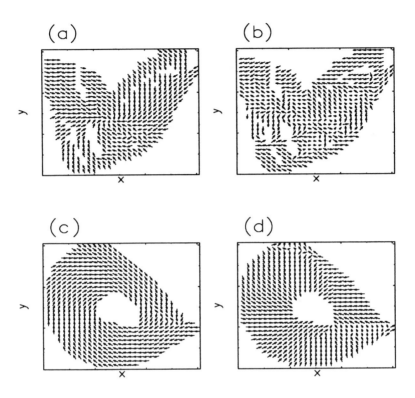

Fig. 1. Directional element fields in dimension $d = 2$ under DCC on x for Ikeda map with delay = 1 and Rössler system with delay = 2 (standard parameter values - see text). Time series length $N = 20,000$ points, grid size $n^2 = 30 \times 30$. (a) Ikeda map with $\{c_r\} = \{-1, 1\}$. (b) Ikeda map with $\{c_r\} = \{2, -5, 3\}$. (c) Rössler system with $\{c_r\} = \{-1, 1\}$. (d) Rössler system with $\{c_r\} = \{2, -5, 3\}$.

If the time series $v(t)$ generates a continuous phase portrait, we expect the $\dot{x}(x_j)$ in a given grid-box all to be sensibly parallel unless the box is large (cf. Figure 1). So $L_j = 1$ should hold for all j; for random noise on the other hand, this isn't so (Figure 2). Also $L_j = 1$ implies $\Lambda = 1$. In the noise case, if summing the $\dot{x}(x_j)$ over box-j generates an uncorrelated random walk, the expected value of L_j is $R_{n_j}^d$. In this case we get $\Lambda \approx 0$, a result actually realized for Lorenz noise[9]. In general this last is too simple. Nonetheless $\Lambda < 1$ should hold in a clear way for noise. For the examples shown, $d = 3$, $N = 20{,}000$, and the grid size $n^3 = 40 \times 40 \times 40$. We use these parameters for the results shown in the rest of the paper.

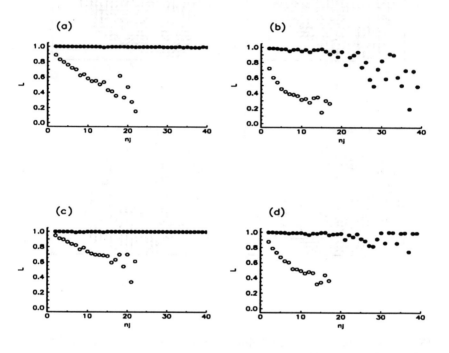

Fig. 2 L_j ploted against n_j for chaotic Lorenz data (solid circles) and Lorenz noise (open circles). In (a) and (b), $\phi = \phi_1(\{c_r\} = \{-1, 1\})$ with (a) $\Delta = 4$ and (b) $\Delta = 7$. In (c) and (d), $\phi = \phi_8(\{c_r\} = \{1, 2, 3, -4, -2\})$ with (c) $\Delta = 4$ and (d) $\Delta = 7$.

To devise a suitable statistical test based on properties of Λ, we must frame suitable hypotheses: (i) the time series $v(t)$ generates continuous phase portraits (H_1), and (ii) $v(t)$ is random (H_0). To this end, we examine properties of Λ in more detail:

(1) For a given choice of vector field, Λ exhibits considerable variation with Δ for both maps and flows (Figure 1). Similarly to what was reported in references[6,11], oscillations of $\Lambda(\Delta)$, although due to finite numerics (grid size, time series length, etc.), are associated with autocorrelation properties of the time series.

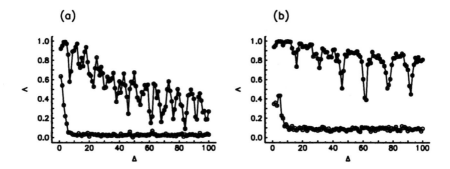

Fig. 3. $\Lambda(\Delta)$ curves for chaotic Lorenz data (upper curves) and Lorenz noise (lower curves). In (a), $\phi = \phi_1(\{c_r\} = \{-1,1\})$. In (b) $\phi = \phi_8(\{c_r\} = \{1,2,3,-4,-2\})$.

Table I. Coefficients c_r for ten vector fields, ϕ_n, used in computations of $\Lambda(\Delta)$

	ϕ_1	ϕ_2	ϕ_3	ϕ_4	ϕ_5	ϕ_6	ϕ_7	ϕ_8	ϕ_9	ϕ_{10}
c_0	-1.0	-3.0	2.0	4.7	-2.0	3.5	-3.4	1.0	0.9	3.0
c_1	1.0	4.0	-5.0	-3.0	3.0	-2.7	-0.5	2.0	0.8	-2.0
c_2	0.0	-1.0	3.0	-1.7	-4.0	-1.4	-0.1	3.0	-3.5	0.0
c_3	0.0	0.0	0.0	0.0	3.0	0.6	4.0	-4.0	4.0	2.0
c_4	0.0	0.0	0.0	0.0	0.0	0.0	0.0	-2.0	-2.2	-3.0

(2) The graph of $\Lambda(\Delta)$ exhibits substantial dependence on the choice of vector field (Figure 3). This is true for both map and flow data, also.

Since $\Lambda = 1$ under H_1 for arbitrary $\{c_r\}$, we regard the choice of vector field as arbitrary. We have computed maximum and minimum values of $\Lambda(\Delta)$ over ten choices of vector field (Table I), for fixed Δ, for smooth time series data and for randomized versions of the same data. Comparisons of plots of maximum and minimum $\Lambda(\Delta)$ values (Figure 4) actually provide us with a simple test. In Figure 4, the two left hand panels (Figures 4(a) and (c)), which

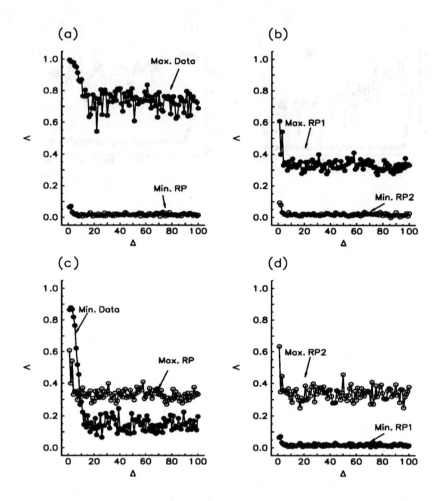

Fig. 4. Ikeda study: maximum and minimum values of Λ over the ten vector fields in Table I for fixed Δ, plotted against Δ, $\Delta = 1,2,...100$. (a) Upper curve: maximum-Λ for Ikeda chaos; lower curve: minimum-Λ for Ikeda noise (RP). (b) Upper curve: maximum-Λ for RP=RP1; lower curve: minimum-Λ for a second independent realization of Ikeda noise (RP2). (c) Upper curve: minimum-Λ for Ikeda chaos; lower curve: maximum-Λ for Ikeda noise (RP). (d) Upper curve: maximum-Λ for RP2; lower curve: minimum-Λ for RP = RP1.

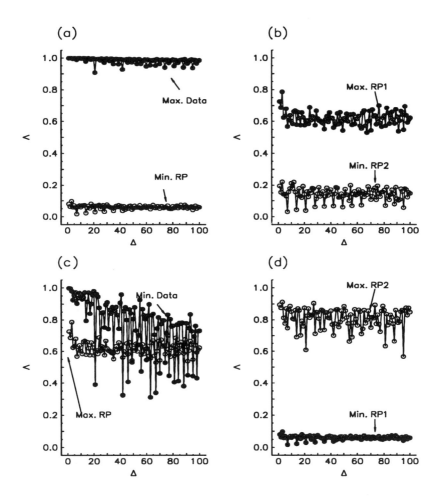

Fig. 5 Rössler study: maximum and minimum values of Λ over the ten vector fields in Table I for fixed Δ, plotted against Δ, $\Delta = 1,2,...100$. (a) Upper curve: maximum Λ for Rössler chaos; lower curve: minimum Λ for Rössler noise (RP). (b) Upper curve: maximum Λ for RP=RP1; lower curve: minimum Λ for a second independent realization of Rössler noise (RP2). (c) Upper curve: minimum Λ for Rössler chaos; lower curve: maximum-Λ for Rössler noise (RP). (d) Upper curve: maximum-Λ for RP2; lower curve: minimum-Λ for RP = RP1.

maximum and minimum $\Lambda(\Delta)$ values (Figure 4) actually provide us with a simple test. In Figure 4, the two left hand panels (Figures 4(a) and (c)), which show maximum and minimum Λ-curves for an Ikeda map time series against (anti-) corresponding Λ-curves for Ikeda noise, are easily distinguishable from one another. The two right hand panels (Figures 4(b) and (d)), which show the same kinds of plots, but with the given data replaced instead by a second, independent realization of Ikeda noise, are, however, indistinguishable from one another. Thus, the given data are not consistent with randomness. In addition, the maximum Λ-curve for the given data (Figure 4(a)) is close to one for all Δ, which is consistent with smoothness.

Thus, taken together, the four panels of Figure 4 tell a clear story. We formulate the statistical decision test as follows:

(1) *use the Λ-statistic in four-panel $\Lambda(\Delta)$-plots as shown herein, in particular noting the requirement for H_1 that $\Lambda \approx 1$, and*

(2) *do this in the setting of a relatively non-restrictive null hypothesis, H_0, where H_0 is the assertion that the given time series is indistinguishable from a suitably randomized version of itself.*

Applied to the Rössler system, this test again gives the correct result (Figure 5). Since the Ikeda spectrum is nearly white, the right hand panels of Figure 4 give us another confirmation of the test for free; viz, that white noise is not smooth. We have applied this test to several additional examples, of Hénon data, an experimental chaotic system, and noise reduced time series, with good results in all cases.

SUMMARY AND DISCUSSION

We have described a statistical test for distinguishing smoothness from randomness for time series data. The method is based on the arbitrariness in the choice of vector field for evaluating a statistic, Λ, which is a natural extension of a quantity first introduced by Kaplan and Glass. Because the test is a simple comparison, the form of Λ in eq. (4) is not likely to be of too much importance. Certainly the random cases studied are not generally consistent with a random walk model for $\dot{x}(x_j)$ since $\Lambda = 0$ is not the rule. But the test is general, that is, it is robust against a variety of behaviors of $\dot{x}(x_j)$ for random data.

ACKNOWLEDGMENTS

LWS wishes to thank Sarah Little for the use of her personal computing facilities, and RC wishes to thank Guan-Hsong Hsu for useful discussions. This work was supported by the Office of Naval Research.

REFERENCES

1. N. H. Packard, J. P. Crutchfield, J. D. Farmer and R. S. Shaw, Phys. Rev. Lett. **45**, 712 (1980).
2. F. Takens, "Detecting Strange Attractors" in Turbulence, in Dynamical Systems and Turbulence, Warwick, 1990, D. A. Rand and L. -S. Young, eds, L. N. Mathematics, No. **898** (Springer-Verlag, Berlin, 1981), pp. 366-381.
3. J.-P. Eckmann and D. Ruelle, Physica D **56**, 185-187 (1992).
4. P. E. Rapp, A. M. Albano, T. I. Schmah and L. A. Farwell, Phys. Rev. E **47**, 2289 (1993)
5. G. Sugihara and R. M. May, Nature **344**, 734 (1990).
6. D.T. Kaplan and L. Glass, Phys. Rev. Lett. **68**, 427 (1992).
7. M.B. Kennel and S. Isabelle, Phys. Rev. A **46**, 3111 (1992).
8. R. Wayland, D. Bromley, D. Pickett and A. Passamante, Phys. Rev. Lett. **70**, 580 (1993).
9. L.W. Salvino and R. Cawley, "Statistical Test for 'Smooth' Dynamics in Embedded Time-Series" (NSWCDD preprint, June 1993).
10. A flow that is only C^0 need not evolve uniquely (deterministically) from a given initial condition. A system with these properties could have nice continuous flow lines and pass all the tests so far advanced in Refs. 5 – 8, as well as our own. A simple one - dimensional example of such a C^0 system is $\dot{x} = (1 - x^2)^{1/2}, x(0) = 0$. (The authers owe this example to G.-H. Hsu.)
11. D.T. Kaplan and L. Glass, Physica D **64**, 431 (1993).
12. The use of the over-dot to denote the directional element $\dot{x}$ has nothing at all to do with time differentiation. It is a purely notational device.

MECHANICAL SOURCES OF CHAOS

CHAOTIC SOURCES OF NOISE IN MACHINE ACOUSTICS

Prof. F. C. MOON
Cornell University, Ithaca, New York

Dipl.-Ing T. BROSCHART
Techn. Hochschule Darmstadt
Federal Republic of Germany

ABSTRACT

In this paper a model is posited for deterministic, random-like noise in machines with sliding rigid parts impacting linear continuous machine structures. Such problems occur in gear transmission systems. A mathematical model is proposed to explain the random-like structure-borne and air-borne noise from such systems when the input is a periodic deterministic excitation of the quasi-rigid impacting parts. An experimental study is presented which supports the model. A thin circular plate is impacted by a chaotically vibrating mass excited by a sinusoidal moving base. The results suggest that the plate vibrations might be predicted by replacing the chaotic vibrating mass with a probabilistic forcing function. Prechaotic vibrations of the impacting mass show classical period doubling phenomena.

INTRODUCTION

A common problem in the design of many machine systems is the elimination of unwanted structurally excited acoustic noise. Examples include computer printers, gear transmissions or speed reducers, and submarines. A common feature of these different systems is an enclosure which radiates the sound into the surrounding environment and internal discrete machine elements such as gears, levers, actuators which excite the structural vibrations of the enclosure. Typically the enclosure is composed of shell or plate-like elements while the internal components can often be modeled as rigid bodies (Figure 1). When the connections between the moving internal rigid-like parts and the shell or plate-like enclosure are linear, then the classical theory of machine acoustics can be used (Cremer and Heckl [2]). In such theories a single frequency excitation inside the machine will produce a single frequency acoustic signal outside the machine. However, if the connection is nonlinear, then a single frequency input can create broadband acoustic noise outside the enclosure. In this paper we will examine this latter class of problems in which deterministic periodic input excitation produces a random-like output. Such problems are now called chaotic vibrations (see, e.g., Moon [10]).

28 Chaotic Sources of Noise

Reprinted from Archive of Applied Mechanics 61

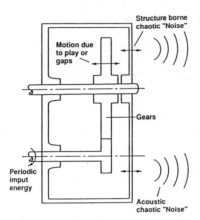

Figure 1. Sketch of multi-body machine elements and elastic enclosure as a machine noise generator.

In this paper we present results of an experimental study of a paradigm for the deterministic generation of broadband machine noise. In particular, the interaction of an elastic plate with a periodically excited impacting mass is presented (Figure 2). Although the plate is considered to be a linear structure, its interaction with the impacting mass is nonlinear because the plate-mass gap creates nonlinear boundary conditions. Another feature of this system is the interaction between a physical system described by a linear partial differential equation and an ordinary differential equation through the equation of motion of the impacting mass. This class of problems has not received much attention in the nonlinear-chaotic dynamics literature.

Often broadband machine noise is treated as a random vibration problem. However, the source of the random excitation is usually assumed with no direct link to deterministic causes. It is the thesis of this paper that many examples of broadband machine noise can be traced to deterministic sources using the new tools of the nonlinear theory of chaotic vibrations. In this way a link can be formed between random vibration of linear structures and nonlinear chaos theory.

In recent years several papers have appeared which attempt to explain chaotic vibrations of machine elements including tool bit chatter, gear transmission rattling, drill bit vibrations, sliding contact problems, and play between moving parts. The key to the explanation of chaotic noise is the identification of nonlinear processes in mechanical systems.

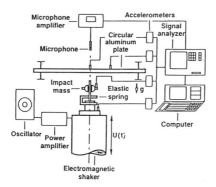

Figure 2. Sketch of experimental apparatus to measure vibrations in a circular plate excited by an impact mass.

Despite the presence of obvious nonlinear effects such as friction, play, and plastic deformation, most acoustic models for machine noise are based on linear theories. In one method of analysis based on a sweeping average over consecutive narrow frequency bands which is suitable for design, the radiated acoustic power $P(\omega)$ at a given frequency ω, is given by (Cremer and Heckl [2, pp. 519-540])

$$P(\omega) = \rho_0 c_0 A h_t^2(\omega) \mathbf{F}^2(\omega) (\sigma(\omega)). \qquad (1)$$

Here ρ_0 and c_0 are the density and velocity of sound in the ambient medium (e.g., air or water), A is the radiating surface area of the machine, $h_t^2(\omega)$ is the square of the mean transfer function for structure borne sound and $\mathbf{F}^2(\omega)$ is the mean exciting force. The term $\sigma(\omega)$ is the radiation efficiency. This formulation has been successfully applied to machines with casings with box-type structures consisting of rectangular plates (Müller et al. [12]). It is also possible to include the effects of ribs attached to the surface of the casing (Richter [16]) and the effect of point masses.

CHAOTIC NOISE DUE TO IMPACT

Among the primary sources of nonlinearities in machine dynamics is play or gaps between moving parts and/or the machine structure. For example, recent research by Pfeiffer and coworkers [13-15] have examined the source of noise in gear transmission systems. In a recent paper Hongler and Streit [5] have shown the relationship of chaotic noise in gear transmissions to impact oscillators. In an impact oscillator, the interaction time between moving parts or a moving part and the machine structure is usually short compared with the period of the exciting frequency. Several simple models of these so-called impact oscillators have been studied and both period doubling subharmonic bifurcations, as well as chaotic vibrations, have been shown to be typical outputs for periodic vibration inputs see e.g., Holmes [4], Lichtenberg and Lieberman [8], Shaw and Holmes [18], Shaw [17], Tufillaro and Albano [20], Isomäkii et al. [6], Li et al. [7]. Dissipation during impact is often assumed in the form of a coefficient of restitution β given by

$$[U_n^+ - \dot{U}(t)] = \beta[U_n^- + \dot{U}(t_n)] \tag{2}$$

where U_n^+ is the post-impact velocity of a freely moving impact mass at time $t = t_n$ directed away from the periodically driven surface, U_n^- is the pre-impact velocity and $\dot{U}(t_n)$ is the velocity of the moving driven surface assumed positive in the direction of the surface normal. In some problems the motion between impacts can be integrated, and a set of difference equations or point map can be derived. An example is the so-called Fermi map with dissipation. This map describes the motion of a mass oscillating between two walls; one is moving with a known periodic motion $U(t)$ while the other is fixed. The discrete variables are the normalized impact time τ_n and the normalized velocity before the n-th impact:

$$\tau_{n+1} = \tau_n + 1/|U_n|, \quad U_{n+1} = \gamma U_n - 2\beta \dot{U}(\tau_{n+1}). \tag{3}$$

Another example is a mass bouncing on one moving surface under gravity. The resulting map is called the standard map and has been studied by many authors. This map is given by the two difference equations:

$$\tau_{n+1} = \tau_n + U_n, \quad U_{n+1} = \gamma U_n - 2\beta \dot{U}(\tau_{n+1}). \tag{4}$$

In both problems the interaction of the moving mass with the impact masses is accounted for in the coefficient of restitution γ and there is no damping between impacts. When linear damping is added between impacts, then the map becomes implicit. In the case of the standard map with linear viscous damping η between bounces, the map is given by

$$\tau_{n+1} - \tau_n = (1/\eta)[1 - e^{-\eta(t')}] + (U_n/g)[1 - e^{-\eta(t')}],$$

$$U_{n+1} = (1+\gamma)\dot{U}(\tau_{n+1}) + \gamma[(g/\eta(e^{-\eta(t')} - 1 + U_n e^{-\eta(t')}] \tag{5}$$

where $t' = \tau_{n+1} - \tau_n$ and g is the acceleration of gravity. These equations, however, differ from (3), (4) in that the unknown τ_{n+1} must be determined implicitly.

The case that we will examine experimentally in this paper combines features of both the Fermi and standard maps, namely a mass vibrating between two walls under gravity with viscous dissipation. This case is shown in Figure 2. To connect this problem to machine acoustics, however, we let one of the impacted surfaces be a linear structure. In our experiment this structure was a thin circular plate.

Thus, our paradigm for nonlinear machine acoustics is as follows: The periodically vibrating base excites a periodic motion of the mass for sufficiently small excitation. As the level of excitation increases, the motion of the mass goes through a series of bifurcations such as the period doubling scenario which leads to chaotic behavior. The plate at first undergoes a series of periodic impulses, and when the mass becomes chaotic, the plate is driven by a set of random-like impulse forces from the moving mass. If one can derive a probability density function for the impulses for both the time of impact and the size of the impact, then one can calculate the radiated sound from the plate using the analytical tools of random vibration theory. It is the thesis of this paper that the probability density function for the structural impulses can be approximated by random phase assumption on the time between impacts and an exponential distribution for the impulse to the plate.

DESCRIPTION OF EXPERIMENT

A sketch of the apparatus is shown in Figure 2. In order to examine a simple model, we decided to choose two simple subsystems whose independent dynamics are well understood. This led to the choice of a circular plate with free edges as the linear acoustic radiating element and an unattached mass bouncing on a vibrating spring element under gravity as the nonlinear chaos oscillator. The geometry, masses, and stiffnesses of the elements are shown in Table 1. The aluminum plate was mounted on four soft rubber supports which resulted in a rigid body mode of 10 Hz. The impacting mass was constrained to one-dimensional, vertical motion by a rod attached to the shaker. The mass was positioned so that impact would occur at the center of the plate. The impact mass on the spring had a linear frequency of 40 Hz. The lowest plate modes were measured up to around 400 Hz and are listed in Table 2. The measured values agreed very well with theoretical values. The lowest flexural mode however, (i.e., two modal diameters, 21 Hz) did not show a strong resonance, perhaps because of the damping of the rubber supports. Several exciting frequencies were studied below 100 Hz, with the principal data taken at 42, 72 Hz.

TABLE 1. Geometry, mass, stiffness of machine elements

Circular plate	material: Aluminum; Diameter: 500 mm; thickness: 1 mm; mass: 530 g
Impact mass	Material: Brass; mass: 24 g
Spring	Stiffness: 3.7 N/mm
Plate supports	Material: Rubber; stiffness: 0.5 N/mm
Accelerometers	Added mass on impact element:2.2 g; Added mass on plate:2.2 g

TABLE 2. Principal frequencies

Impact mass on shaker spring	60 Hz
Rigid body plate mode on 4 rubber supports	10 Hz

Measured elastic plate modes
  symmetric modes (Hz)
  nodal diameter modes (Hz)
	(n = no. nodal diameters, s = no. nodal circles)
	(0,1) = 37, (0,2) = 158, (0,3) = 356;
	(2,0) = 21, (3,0) = 49, (4,0) = 93, (5,0) = 132, (6,0) = 185, (7,0) = 248, (1,1) = 87, (2,1) = 146, (3,1) = 214, (4,1) = 298, (1,2) = 239

Shaker input frequencies (Hz)	42,72

The initial gap between the bottom of the plate and the top of the impacting mass was varied from 2 to 6 mm, with the principal data taken at 2.5 mm. The excitation amplitude of the vibrator was around 2 mm peak to peak.

The procedure was to excite the shaker at a fixed frequency until the impact mass began to bounce on the spring attached to the shaker. The amplitude of the shaker was then increased until the mass began hitting the plate.

Vibration measurements were taken with Bruel and Kjaer accelerometers attached to the shaker, the impacting mass, and the circular plate. In some experiments, two accelerometers were placed on the plate to measure cross-correlation between signals at different locations on the plate. In addition, a B & K acoustic microphone was mounted above the plate.

The data was analyzed in a 4-channel signal analyzer (Scientific Atlanta) using fast Fourier transforms (FFT), autocorrelation, cross-correlation, and probability density functions.

EXPERIMENTAL RESULTS

Preliminary experiments were carried out on the two subsystems, the circular plate and the free mass-spring shaker system. The natural frequencies of the plate were measured using both transient impact FFT measurements as well as exciting the Chladni patterns in the plate with an acoustic sound generator. These experiments showed that the four rubber supports did not distort the free-free plate vibrations which checked quite closely the theoretical natural frequencies.

The bouncing ball on a spring is known to undergo period doubling oscillations. Independent experiments were carried out without the plate, and period two and period four subharmonic oscillations were observed (Figures 3a,3b).

Period three motions of the bouncing mass are shown in Figure 3c along with the autocorrelation function for this case (Figure 3d). Notice that the autocorrelation function has the same periodicity as the time history of the mass.

Data for chaotic vibrations of the impact mass-plate system are shown in Figures 4a-4d. Figures 4a-4c show the broadband noise generated by a 42 Hz excitation of the shaker. However, the circular plate shows a peak at the second symmetric mode, (i.e., one modal diameter). In comparison with the periodic motion in Figure 3d, the autocorrelation function for the impact mass (Figure 4d) shows a decay in time indicating that events greater than two hundreds of a second are not correlated very well. Unity divided by this decay time is thought by some to be a good measure of the Lyapunov exponent.

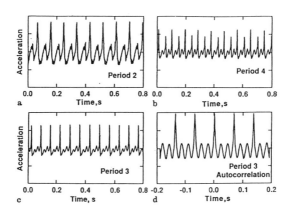

Figures 3a-3d. Period doubling and other subharmonic motions of the periodically excited impact mass (excitation frequency: 42 Hz); (a) period-two motions, (b) period-four motions, (c) period-three motions, (d) autocorrelation function for period-three motions.

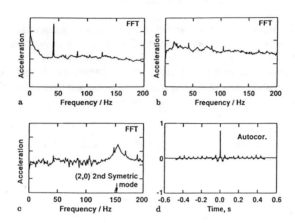

Figures 4a-4d. Chaotic vibrations of a mass impacting a circular plate at its center (excitation frequency: 42 Hz); (a) FFT of electromagnetic shaker, (b) FFT of accelerometer output from moving mass, (c) FFT of accelerometer output at plate center, (d) autocorrelation function of impacting mass

Probability density functions (PDF) for the chaotic vibrations in Figure 4 are shown in Figures 5a to d. These data are, of course, averaged over many time records, but they show a remarkable smoothness which suggests that there may be an underlying theory from which one could calculate these functions as by Tsang and Liebermann [19]. One should notice that the PDF of the shaker shows a characteristic shape for a periodic sinusoidal function. The PDF for the plate acceleration, however (Figures 5c, 5d), shows a symmetric exponential distribution. The PDF for the impact mass (Figure 5b), however, shows a skewed probability distribution.

Of particular interest was the cross-correlation function (CCF) between acceleration signals at different positions in the plate. Figure 6 shows a comparison of CCF for one accelerometer located at the plate center and a second accelerometer located at some radial distance from the center. These data seem to indicate that as one moves away from the impact point, the motion becomes spacially uncorrelated. This might suggest a kind of spacial chaos in the plate.

F. C. Moon and T. Broschart 35

Reprinted from Archive of Applied Mechanics 61

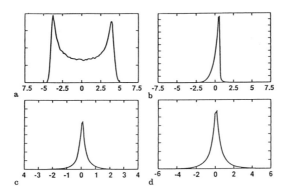

Figures 5a-5d. Probability distribution functions for chaotic vibrations in Figure 4. (a) vibration shaker, (b) impact mass acceleration, (c) plate acceleration at center, (d) plate edge acceleration.

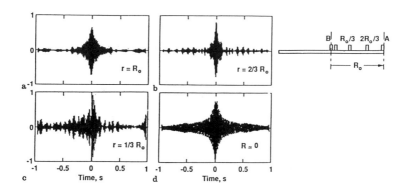

Figure 6. Cross-correlation function for chaotic vibrations in Figure 4 (both signals taken from accelerometers on the plate, one at the center and the other at various radii along the plate)

Finally, in Figure 7 we show a curve fit for the PDF in Figure 5c of the vibrations at the plate center which shows that an exponential distribution function $P_0 \exp(-g|x|)$ seems to provide a good approximation to the experimental data. This suggests that one might try to replace the deterministic impact mass with a sequence of impact forces at the plate center with random impact times and with an exponential distribution of intensities in order to determine the statistics of the plate motion. Thus, an attempt to find probability distribution functions for deterministic chaos problems may provide a link to the techniques of random vibration methods used in linear systems.

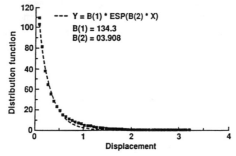

Figure 7. Exponential interpolation of probability distribution function of chaotic vibrations at the center of the plate for excitation frequency of 42 Hz (------exponential curve fit; ■ experimental data).

DISCUSSION

The experimental results presented above show that the interaction between a linear structural acoustic radiator and a deterministic nonlinear rattling device can produce vibrations in the linear structure that look like a random vibration problem. This suggests the following proposal. Can one replace the deterministic source of chaotic vibrations with a random process such that the resulting structural and acoustic vibrations exhibit the same statistics as the deterministic chaos? If this is possible, then the corollary offers even more promise, namely the possibility of finding deterministic models for what is now considered random noise.

The use of probability measures in rattling dynamics has already been mentioned in the introduction by Tsang and Liebermann [19], and Pfeiffer and Kücükay [15]. The technique has also been used for the Lorenz equations where the three first order deterministic equations for motion are replaced by a second order system with stochastic inputs. Tsang and Lieberman [19] replace the equation for the phase of the impact, or the time of impact, by a random phase assumption. This suggests a similar method for the rattling problem discussed in the experimental results. Everson [3] replaces the phase equation by a random phase assumption (bouncing ball problems) in another paper.

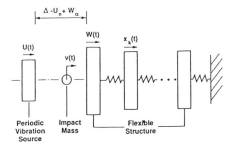

Figure 8. Sketch of particle mass impact with multi-degree of freedom structure as a simple model for generation of machine noise

To illustrate these ideas, we present an idealized model or paradigm for the type of experiments we have presented above. We consider a mass which impacts and oscillates between two moving walls as shown in Figure 8. The left wall is assumed to oscillate with a known periodic motion U(t), while the right mass is assumed to be part of a linear structure with N degrees of freedom. This model differs from our experiments in that we do not include gravity in the analysis.

To construct a model for the impact dynamics, one needs an equation of motion for the time between impacts and impact laws at the right and left walls. In addition, one must make some assumption about multiple impacts at each wall before velocity reversal. The velocity of the impacting particle is denoted by v(t).

We first define the impact times at the left wall with the sequence $\{t_n\}$, and those at the right wall by $\{t_\alpha\}$. The impact conditions at each wall assume that the post impact relative velocities of the mass is a fraction of the preimpact relative velocity, i.e.

$$V_n^+ - \dot{U}_n = \beta_1(V_n^- + \dot{U}_n) \quad \text{at the left wall,} \qquad (6)$$

$$V_\alpha^+ + \dot{W}_\alpha^+ = \beta_2(V_\alpha^- - \dot{W}_\alpha^-) \quad \text{at the right wall,} \qquad (7)$$

$$\text{where } v(t_n^-) = -V_n^-, \ v(t_n^+) = V_n^+, \qquad (8)$$

$$\text{and } v(t_\alpha^-) = V_\alpha^-, \ v(t_\alpha^+) = -V_\alpha^+ \qquad (9)$$

and $\dot{U}(t_n) = \dot{U}_n$ is the velocity of the left wall at impact, and $\dot{W}(t_\alpha^+) = \dot{W}_\alpha^+$, $\dot{W}(t_\alpha^-) = \dot{W}_\alpha^-$ are the post- and pre-impact velocities of the right hand wall.

We assume here that the left wall has sufficient mass to be unaffected by the mass impact. However, at the right wall we assume that the linear momentum is conserved during the impact between the mass and the mass element in the structure.

One further assumption is that the time of impact is so small that only the impacted mass in the many-body linear system participates in the momentum conservation law. This condition takes the form

$$mV_\alpha^- + M\dot{W}_\alpha^- = -mV_\alpha^+ + M\dot{W}_\alpha^+ \qquad (10)$$

where m is the impacting mass and M is the mass of the wall.

The impact equations (6), (7) involve seven velocities: $\dot{U}_n$, V_n^-, V_n^+, V_α^-, V_α^+, $\dot{W}_\alpha^-$, $\dot{W}_\alpha^+$. We have already assumed that $\dot{U}(t_n^-) = \dot{U}(t_n^+) \equiv \dot{U}_n$. If the impacts alternate between left and right walls, then the integration of the equation of motion, neglecting gravity, leads to the ancillary conditions when the impact times are ordered as $\{t_n, t_\alpha, t_{n+1}\}$:

$$V_n^+ = V_\alpha^-, \qquad V_\alpha^+ = V_{n+1}^- \qquad (11)$$

Problems arise when multiple impacts with one wall occur (see e.g., Bapat et al. [1]).

For alternating impact the impact times t_α, t_n are determined by

$$t_\alpha = t_n + \{(\Delta + W_\alpha - U_n)/|V_n^+|\} \qquad (12)$$

$$t_{n+1} = t_\alpha + \{(\Delta + W_\alpha - U_{n+1})/|V_\alpha^+|\} \qquad (13)$$

If the impacts are indeed alternating, the absolute value signs are superfluous, but in calculations one may encounter negative V_n^+, V_α^+.

In addition to the impact equations, one needs equations of motion for the linear structure between impacts $t_\alpha \leq t \leq t_{\alpha+1}$. The dynamics $x_k(t)$ of all the structural masses are given as free vibrations of the system determined by the initial conditions $x_k(t_\alpha)$, $\dot{x}_k(t_\alpha)$. For simplicity, let us assume that $x_1(t)$ is the motion of the left wall. Then we assume that

$$x_k(t_\alpha^-) = x_k(t_\alpha^+), \quad \dot{x}_1(t_\alpha^-) = \dot{W}_\alpha^-,$$

$$\dot{x}_1(t_\alpha^+) = \dot{W}_\alpha^+, \dot{x}_k(t_\alpha^-) = \dot{x}_k(t_\alpha^+) \quad (k = 2,3, ..,N) \qquad (14)$$

Thus, we assume that the displacements before and after impact are continuous. Also, the velocity of all masses except $\dot{x}_1$ are continuous. The jump in velocity $\dot{x}_1(t_\alpha^-) - \dot{x}_1(t_\alpha^+)$ is given by the impact and momentum conservation equations (7)-(10). The equations of motion of the linear structure are given in the standard matrix format

$$M\ddot{x}(t) + Kx = 0. \qquad (15)$$

The general solution may be found in standard texts such as Meirovitch [9]. For the single degree of freedom case without damping, the dynamics of the right wall between impacts is given by

$$W_{\alpha+1} = W_\alpha \cos[\yen] + (\dot{W}_\alpha^+/\Omega) \sin[\yen],$$

$$\dot{W}_{\alpha+1}^- = -\Omega W_\alpha \sin[\yen] + \dot{W}_\alpha^+ \cos[\yen] \tag{16}$$

where $\yen = \Omega(t_{\alpha+1} - t_\alpha)$
and Ω is the eigenfrequency of the wall oscillator.

Multiple impact on the same wall is a difficult problem and generally results in an implicit set of equations (see e.g., Bapat et al. [1]). Some authors, however, have adopted an artificial device of only using the absolute value of the post impact velocity in order to insure positive time increments and alternating impact. This assumption was used in this study also.

As an example of the procedure outlined above for an analytical model of deterministic generation of rattling machine noise, we present only a few numerical results for impact of a one degree of freedom oscillator. Further analysis of this model is under study.

To obtain a set of difference equations for a sinusoidal wall motion $U(t) = U_0 \sin(\omega t)$ we introduce the nondimensional variables

$$S_n = \omega t_n, \quad T_\alpha = \omega t_\alpha, \quad D_{\alpha+1} = T_{\alpha+1} - T_\alpha,$$

$$V_n = V_n^+/(\omega\Delta), \quad F_n = V_\alpha^+/(\omega\Delta)$$

$$X_\alpha = W_\alpha/\Delta, \quad Y_a = \dot{W}_\alpha^-/(\Delta\omega), \quad Z_\alpha = \dot{W}_\alpha^+/(\Delta\omega) \tag{17}$$

Finally, we assume that the wall motions are small compared with the gap Δ, i.e., $|U| \ll \Delta$, $|W| \ll \Delta$. This assumption converts the implicit equation into an explicit map. If we add structural damping γ to the single degree of freedom oscillator, this map has the form

$$S_{n+1} = T_\alpha + (1/|F_\alpha|),$$

$$V_{n+1} = \beta_1 F_\alpha - (1+\beta_1) A \sin(S_{n+1}),$$

$$D_{\alpha+1} = 1/|V_{n+1}| + 1/|F_\alpha|,$$

$$Y_{\alpha+1} = e^{-\gamma D_{\alpha+1}}[Z_\alpha \cos(\rho D_{\alpha+1}) - (\rho X_\alpha + \gamma/\rho)(Z_\alpha + \gamma X_\alpha)\sin(\rho D_\alpha D_{\alpha+1})],$$

$$F_{\alpha+1} = [\beta_2(V_{n+1} - Y_{\alpha+1}) - \eta\, V_{n+1} - Y_{\alpha+1}] / (1+\eta)$$

$$Z_{\alpha+1} = [\eta\, V_{n+1} + Y_{\alpha+1} + \eta\, \beta_2(V_{n+1} - Y_{\alpha+1})] / (1+\eta)$$

$$X_{\alpha+1} = e^{-\gamma D_{\alpha+1}}[X_\alpha \cos(\rho D_{\alpha+1}) + (Z_\alpha + \gamma X_\alpha)(1/\rho)\sin(\rho D_\alpha D_{\alpha+1})],$$

$$T_{\alpha+1} = T_\alpha + D_{\alpha+1} \qquad (18)$$

where S_n, T_α are mod (2π) variables. In these equations β_1, β_2 represent energy loss at impact (6), (7), while γ represents damping in the structural oscillator. The ratio of the natural frequency to forcing frequency is denoted by $\Omega/\omega = \rho$, while the impact mass to oscillator mass is given by $c = m/M$. The nondimensional forcing amplitude is represented by A.

NUMERICAL RESULTS

The Equations (18) were programmed to run as a set of coupled difference equations using the following parameters values:

   mass ratio $c = m/M = 0.1$,
   forcing amplitude $A = 0.2$,
   natural frequency to forcing frequency is $\rho = \Omega/\omega = \pi$,
   impact loss coefficients $\beta_1 = \beta_2 = 0.8$,
   oscillator damping $\gamma = 0.05$.

This set of parameters seems to lead to chaotic looking dynamics. The results for 1000 iterates plotted as a probability density function $P(X)$ are shown in Figure 9 where 100 bins were set up as an approximation to $P(X)$. The amplitude range for the oscillator amplitude X_n was $-0.05 \leq X \leq 0.05$. We can see that the distribution looks like the exponential function in Figure 7 for the experimental data from the impact of the circular plate. The maximum value of X for up to 4000 iterates was 0.97×10^{-2}.

The similarities between this simple extension of the Fermi rattling oscillator (3) represented by (18) and the experimental impact of a continuous plate give encouragement that simple models may be found to provide insight into the deterministic sources of machine noise. However, the map (18) represents a fourth order map which has much greater complexity than most of the chaos maps in the applied mathematics literature. Hence, further analysis of (18) must await future research reports.

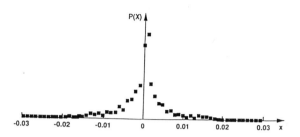

Figure 9. Probability density function for plate displacement from numerical calculation of theoretical impact map (18)

ACKNOWLEDGMENTS

The first author wishes to thank the Federal Republic of Germany for providing support for this research through an Alexander von Humboldt Foundation Senior Scientist Award during the period August 1988 through January 1989. This research could not have been undertaken without the support and inspiration of Prof. Dr.-Ing. F. Kollmann of the Technische Hochschule Darmstadt in whose laboratory these experiments where carried out. Professor Kollmann also provided some background material for this report on linear machine acoustics. Thanks is also due to Prof. Dr.-Ing. H. Wölfel in Darmstadt who, along with Prof. Kollmann, sponsored the first authors Humboldt visit to Germany.

REFERENCES

1. Bapat, C. N.; Sankar, S.; PuppleWell, N.: Repeated impacts on a sinusoidally vibrating table reappraised. J. Sound Vibration 108 (1986) 99-115
2. Cremer. L.; Heckl, M.: Structure-borne sound (2nd ed.). New York: Springer 1988
3. Everson, R.M.: Chaotic dynamics of a bouncing ball. Physica I9D (1986) 355–383
4. Holmes, P. J.: The dynamics of repeated impacts with a sinusoidally vibrating table. J. Sound Vibration 84 (1982) 173 -189
5. Hongler, M.O.; Streit, L.: On the origin of chaos in gearbox models. Physica 29D (1988) 402-408
6. Isomäki, H. M.; Van Boehm, J.; Räty, R.: Fractal basin boundaries of an impacting particle. Physics Letters A 126 (1988) 484-490
7. Li, G.X.; Rand, R. H.; Moon, F. C.: Bifurcations and chaos in a forced zero-stiffness impact oscillator. Int. J. Non-Linear Mechanics 25 (1990) 417
8. Lichtenberg, A. J.; Liebermann, M. A.: Regular and stochastic motion. Berlin: Springer I983
9. Meirovitch, L.; Analytical methods in vibrations. New York; MacMillan 1967
10. Moon, F. C.: Chaotic vibrations. New York: J. Wiley & Sons 1987
11. Moon, F. C.; Shaw, S. W.: Chaotic vibrations of a beam with nonlinear boundary conditions. J. Nonlinear Mech. 18 (1983) 465–477
12. Müller, H.; Langer, W.; Richter, H.P.; Storm. R.: Praisreport Maschinenakustik-Berechlnlungs und Abschätzverfahren für Maschinengeräusche. Forschungskuratorium Maschinenbau e. V., Heft 102, Frankfurt 1983
13. Pfeiffer, F.; Seltsame Attraktoren in Zahnradgetrieben. Ing. Arch. 58 (1988) 113-125
14. Pfeiffer, F.; Kücükav, F.: Eine erweiterte mechanische StoBtheorie und ihre Anw endung in der Getriebedvnamik. VDI-Z. 127 (1985) 341
15. Pfeiffer, F.; Kücükay, F.: Über Rasselschwingungen in Kfz-Schaltgetrieben. Ing. Arch. 56 (1986) 25-37
16. Richter, H.-P.: Maschinenakustische Berechnungen mit dem Programmsystem MASAK (MASchinenAK;ustik), VDI-Bericht Nr. 629, pp. 23-39. Düsseldorf: VDIVerlag 1987
17. Shaw, S. W.: The dynamics of a harmonically excited system having rigid amplitude constraints (Parts 1 and 2). Trans. ASME/J. Appl. Mech. 52 (1985) 453-464
18. Shaw, ,S. W.; Holmes, P. J.: A periodically forced piecewise linear oscillator. J. Sound Vibration 90) (1983) 129 - 155
19. Tsang, K.Y.; Lieberman, M. A.: Analytical calculation of invariant distribution of strange attractors. Physica 11D (1984) 147-166
20. Tufillaro, N. B.; Albano, A. M.: Chaotic dynamics of a bouncing ball. Am. J. Phys. 54 (1986) 939–944

Chaos in gearbox vibrations

Ted Frison

Randle, Inc.
P.O. Box 1010, Great Falls, Virginia 22066
(702) 759-5257

ABSTRACT

We use the methods reviewed by Professor Abarbanel[1] to show that accelerometer data from a gearbox are chaotic. These data have broad-band Fourier components and a natural question arises as to whether these data are chaotic. The requirements for chaos are that:

The Fourier spectrum is broad-band,
There is a non-integer fractal dimension,
There is at least one positive Lyapunov exponent.

These analyses are a prelude to fault prediction and analysis. In this paper we present the results of our studies on a good high-speed gearbox data. That is, the gearbox was known to contain no faults.

2. INTRODUCTION

Predictive fault analysis of rotating machinery has proven to be a difficult problem. Before taking on the problem of fault prediction and analysis, however, there is a natural curiosity about the nature of these systems and their signals. The insights derived from the analysis of baseline data provide guidance for the future development of maintenance tools.

The methods are reviewed in detail in [1]. The general idea is to reconstruct the signal's underlying attractor using phase space reconstruction[2]. Once the attractor has been reconstructed, its invariant geometric and information propagating properties are used to classify and analyze the behavior of the system.

The configuration of the gearbox is shown in figure 1. The input spur pinion driven by a high-speed turbine. For this test, the developmental gearbox was statically mounted on a test bench. The signal trace of an accelerometer mounted on the case is shown in figure 2, along with its FFT. The sensor, an accelerometer, was sampled at 320 khz and low-pass filtered at 80 khz. The data were stored on a modified video recorder. These data are especially interesting because they were collected under circumstances similar to the shop conditions that will exist for the final application. Thus, they contain all the artifacts that one would expect from real data.

3. CHAOTIC CHARACTERISTICS OF A HIGH-SPEED GEARBOX

The main menu for the chaotic signal processing ("CSP") system used in these studies is shown in figure 3. The modules that perform the calculations are platform independent and run on Sun workstations, Apple Quadras, Cray supercomputers, and even an occasional IBM PC. The status windows for four processes running in the background are in the upper left corner.

44 Chaos in Gearbox Vibrations

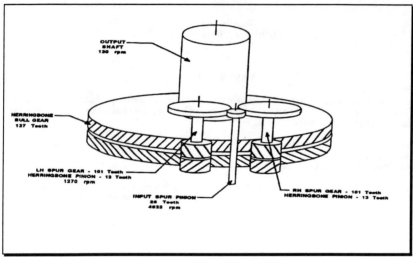

Figure 1. Schematic of gearbox.

3.1. Find the time delay for phase space reconstruction.

Average mutual information, $I(T)$, is a prescription for selecting an appropriate time delay interval (T) for reconstruction of the attractor [3] and is the amount of knowledge (expressed as bits) that one can derive about two datums separated by the time delay, T. It is defined in two dimensions as the joint probabilities of the two datums:

$$I_{AB}(T) = \sum P_{A,B}(a,b) \log_2 \left[\frac{P_{A,B}(a,b)}{P_A(a)P_B(b)} \right]$$

The determination of the time lag, T, is important because an optimum selection of T gives best separation of neighboring trajectories within the minimum embedding space. This is important because calculation of the Lyapunov exponents relies on solving a matrix that is comprised of descriptions of how close trajectories diverge. If the trajectories are not separated, then the matrix will be ill conditioned and may not be solvable.

In the reconstructed attractor, if the time delay is too small, there is little new information contained in each subsequent datum. If T is too large, $x(n)$ and $x(n+T)$ will appear to be random with respect to each other for a chaotic system. The first local minimum of $I(T)$ determines an optimum value of T. Past this point in a chaotic system, ambiguities in the correlation between $x(n)$ and $x(n + T)$ arise -- they start to appear random with respect to each other. The state space portrait begins to lose resolution and the fractal nature of the attractor starts to become blurred.

The average mutual information for the baseline (known good) gearbox data is shown in figure 4. The first local minima, at sample 6 (1.875x10⁻⁵ sec), is a shallow minima. The higher peaks correspond to events that are occurring on an average of 149.7 samples. These events are the engagements of the input spur pinion and the intermediate spur gears. The rate of engagement is 2,137,66 meshings per second. These artifacts are evident across the entire data collection, as are the lower mesh rates from the teeth on the other gears.

3.2. Find the minimum embedding dimension.

The full behavior of a system described by n independent variables can be observed in an n dimensional "state" space. However, the attractor of the system may be contained in a subset of the state space with dimension d_A, and may be described in a state space, d, that is much smaller than n. This minimum embedding dimension d_E is, at most, the first integer greater than $2d_A$ (the correlation dimension); it may be less.

Determination of the minimum embedding dimension, d_E, is of practical interest because the computation burden rises dramatically as dimension increases. Further, noise fills all dimensions, so computations carried out in a higher-than-necessary dimension will be corrupted by noise. If d_E is too small, the trajectory may cross itself., Neighbors near the crossing may be indistinguishable in lower dimensions.

An easy method of determining minimum embedding dimension is used in our processor[4]. As dimension is increased, attractors "unfold." Points on trajectories that appear close in dimension d may move to a distant region of the attractor in dimension $d+1$. These are "false" neighbors in d and the method measures the percentage of false neighbors as d increases. Trajectories that are close in d are tallied, and the number of these trajectories that become widely separated in $d+1$ are calculated. Over the data set one tallies

$$\left[\frac{R_{d+1}^2(n,r) - R_d^2(n,r)}{R_d^2(n,r)}\right] = \frac{|x(n+Td) - x^{(r)}(n+Td)|}{R_d(n,r)} > R_{tol}$$

where R_d is the Euclidian distance between a point and its nearest neighbor and R_{tol} is the criteria for declaring whether the neighbors that are close in d are distant in d+1.

A second criteria is necessary because the nearest neighbor is not necessarily "close." If the nearest neighbor to a point is false but not close, then the Euclidian distance in going to d+1 will be $\approx 2R_a$. So, the second criteria is

$$\frac{R_{d+1}(n)}{R_A} > R_{tol}$$

where

$$R_A^2 = \frac{1}{N}\sum_{n-1}^{N}[x(n) - \bar{x}]^2$$

A nearest neighbor is false if either test fails.

Figure 5 shows the global false nearest neighbor calculations for these data. The data are clearly higher dimension, which is not surprising considering that the accelerometer is capturing multipath signals from six gears with 381 teeth. The teeth on the input spur pinion are engaging at a rate of 2,138 engagements per second. The bull gear mesh rate is only 275 engagements per second on each of the two pinions, but it takes almost a half a second for one complete revolution. It would appear that the global embedding dimension for this signal is at least 12, but could be higher.

3.3. Compute the correlation dimension of the attractor.

The fractal dimension of the attractor[5], d_a, provides information on how much of the state space is filled by the system. One interpretation of d_a is that it measures how many degrees of freedom are significant. Another interpretation of the fractal dimension is that it provides a measure of how an object's bulk scales with its size: bulk = size^{d_a}. Bulk that can be associated with volume and size is then interpreted as Euclidean distance. A plane, for example, has dimension two because the area = d^2. The fractal dimension of the attractor, d_a, may be estimated using Ruelle's approach by calculating the number of spheres or boxes, $N(r)$, of size r that capture all points as r approaches zero:

$$d_a \approx \frac{\log(N(r))}{\log(1/r)} \quad as\ r \to 0$$

Grassberger and Procaccia[6] defined a relatively easy approximation, the correlation dimension, that may be done on a PC for high SNR signals. One major issue is the sensitivity of these calculations to signal SNR. The amount of data required to do the calculations may dramatically increase as SNR decreases.

Using 120,000 samples (a little less than 1/2 second) the attractor is shown to be well populated by the histogram of $N(r)$ -- the top curves in figure 6. The bottom curves in figure 6 are the estimates of d_a for all locations along the top curves. The pointer shows where the final estimates are selected, and presented in figure 7. The final estimate of $d_a = 5.9$ is consistent with the estimate of $d_e >= 12$.

3.5. Compute the Lyapunov exponents.

The Lyapunov exponents describe the rate at which close points in the state space diverge. There is one exponent for each dimension. If the Lyapunov exponents are all zero or negative, the trajectories do not diverge and the system is stable. If one or more Lyapunov exponents is positive, the system is chaotic[7]. The Lyapunov exponents are invariant with respect to initial conditions. Therefore, they are another way of classifying a chaotic system. The more exponents one can correctly find, the more accurate predictions of system behavior will be[8].

All the Lyapunov exponents may be calculated from the Jacobian of the map by the QR decomposition technique discussed by "EKRC."[9]

As defined above, Lyapunov exponents are a global invariant because they describe the effect of infinitesimal perturbations over infinite time. Recent approaches examine how perturbations grow in finite time[10] and how these local Lyapunov exponents relate to predictability[11]. The local Lyapunov exponents measure the divergence of trajectories in different regions of state space.

Our calculations of the global Lyapunov exponents in twelve dimensions show that most, if not all, of the exponents are positive.

4. CONCLUSION

These data meet the rigorous definition of chaos: broad-band spectra, a non-integer correlation dimension d_a < d_e and at least one positive Lyapunov exponent. Our curiosity about the global nature of this physical system is satisfied. The next challenge is to translate this knowledge into methods for fault prediction.

5. ACKNOWLEDGMENTS

This work was performed as part of ongoing joint research between Randle, Inc., the Institute for Nonlinear Science, University of California (San Diego), and the gearbox developer. The author thanks Henry Abarbanel for his advice andencouragement.

6. REFERENCES

1. Henry D. I. Abarbanel, Reggie Brown, J.J. Sidorowich, and Lev. Sh. Tsimring, "The analysis of observed chaotic data in physical systems," accepted, *Reviews of Modern Physics*.

2. Eckmann, J.-P. and D. Ruelle, "Ergodic theory of chaotic and strange attractors", *Rev. Mod. Phys.* **57** 3, pp. 617-656, 1985.

3. A. M. Fraser and H. L. Swinney. "Independent coordinates for strange attractors from mutual information," *Phys. Rev. A*, 33:1134-1140, Feb. 1986.

4. Kennel, Matthew B., R. Brown, and H. D. I. Abarbanel, "Determining embedding dimension for phase-space reconstruction using a geometrical construction," *Phy. Rev. A* **45** pp. 3403-3411, 15 March 1992.

5. F. Hausdorff, "Dimension and ausseres Mass," *Math. Annalen*, vol 79, pp. 157-179, 1918.

6. P. Grassberger and I. Procaccia, "Measuring the Strangeness of Strange Attractors", *Physica* **9D**, pp. 189-208, 1983.

7. J.-P. Eckmann and D. Ruelle, *Rev. Mod. Phys.* **57**, 617 (1985).

8. Abarbanel, Henry D. I., "Determining the Lyapunov Spectrum of a Dynamical System from Observed Data", Presented at the SIAM Conference on Dynamical Systems, Orlando, Florida, 8 May 1990.

9. Eckmann, J.P., S.O. Kamphorst, D. Ruelle, and S. Ciliberto, *Phys. Rev.* **A34**, 4971 (1986).

10. Abarbanel, H. D. I., R. Brown, and Matthew Kennel, "Variation of Lyapunov Exponents on a Strange Attractor", *Journal of Nonlinear Science*, **1**, pp. 175-199, 1991.

11. Abarbanel, H. D. I., R. Brown, and Matthew Kennel, "Local Lyapunov Exponents Computed from Observed Data",*Journal of Nonlinear Science*, Vol 2, pp. 343-365, Sept. 1992.

48 Chaos in Gearbox Vibrations

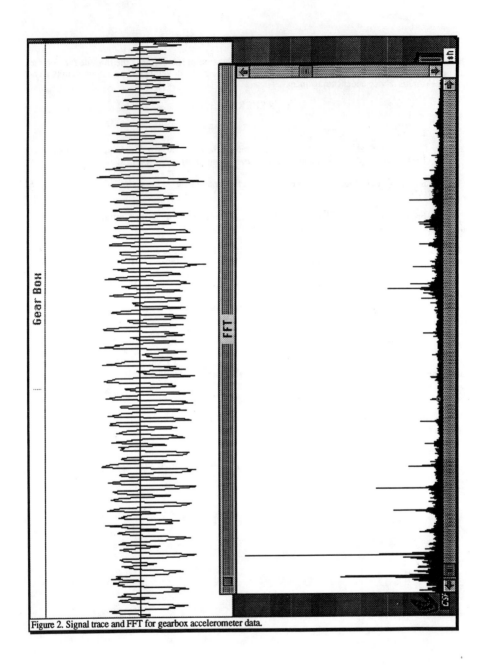

Figure 2. Signal trace and FFT for gearbox accelerometer data.

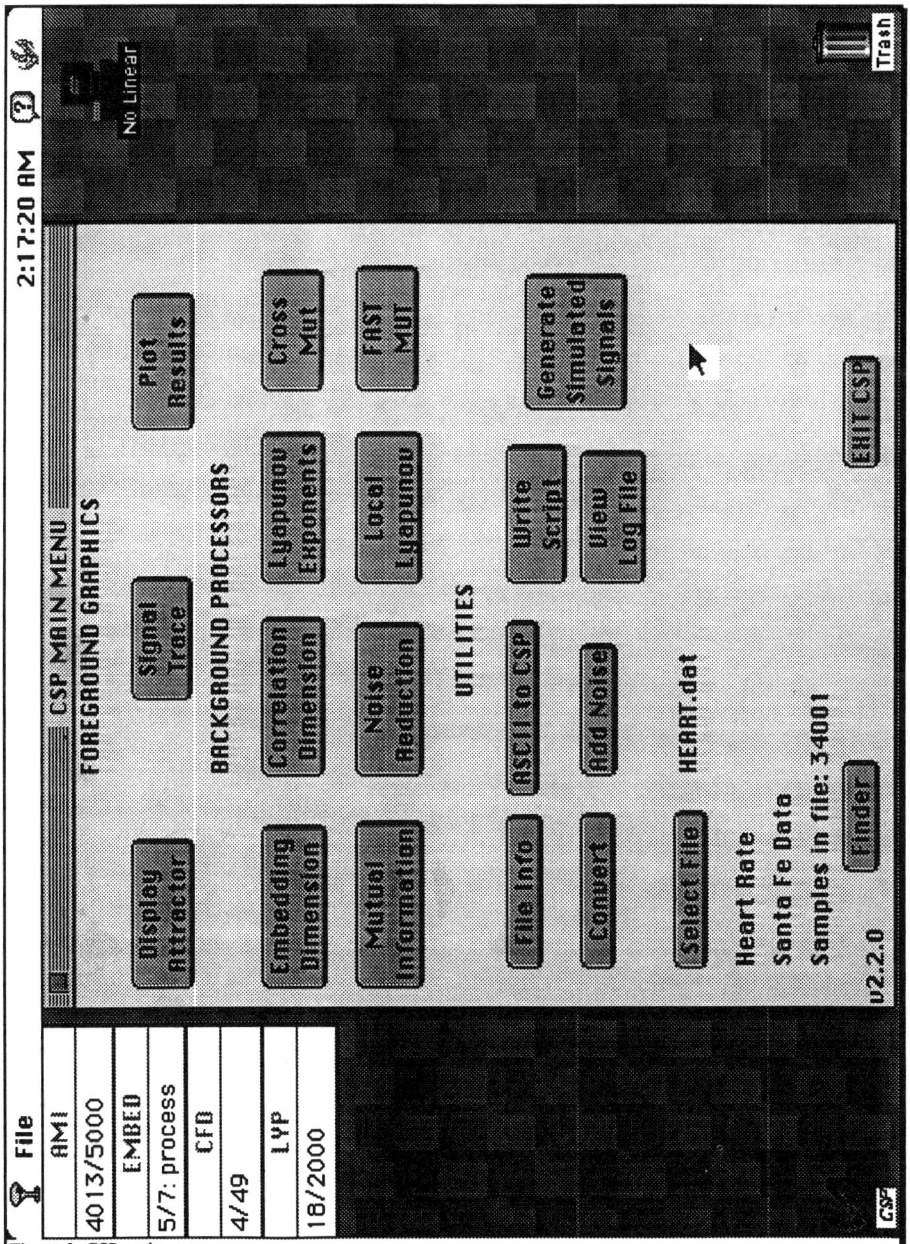

Figure 3. CSP main menu.

50 Chaos in Gearbox Vibrations

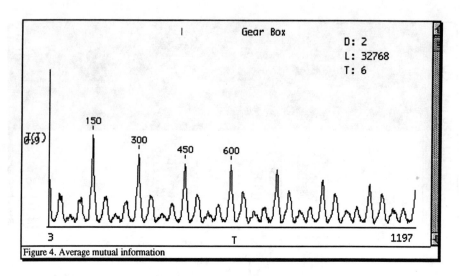

Figure 4. Average mutual information

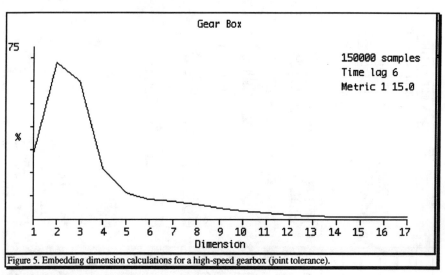

Figure 5. Embedding dimension calculations for a high-speed gearbox (joint tolerance).

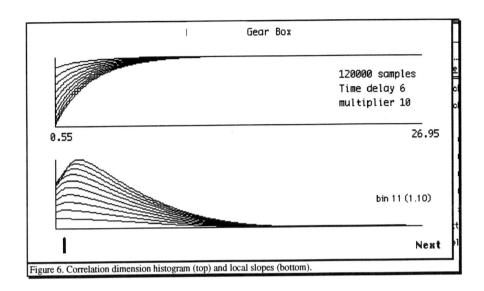

Figure 6. Correlation dimension histogram (top) and local slopes (bottom).

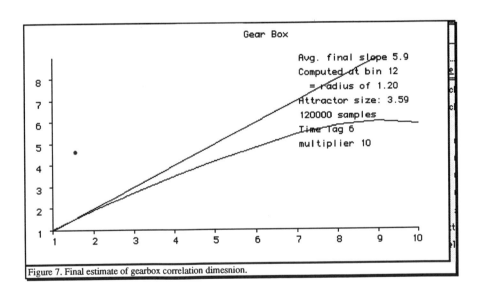

Figure 7. Final estimate of gearbox correlation dimesnion.

TURBULENCE AND CHAOS

Nonlinear Analysis of High Reynolds Number Flows over a Buoyant Axisymmetric Body

Henry D. I. Abarbanel[1],
Department of Physics
and
Marine Physical Laboratory
Scripps Institution of Oceanography
University of California, San Diego
Mail Code 0402
La Jolla, CA 92093-0402

Richard A. Katz,
Naval Undersea Warfare Center Detachment New London
Code 33A, Building 2
New London, CT 06320

Thomas Galib
and
Joan Cembrola,
Naval Undersea Warfare Center Division Newport
1176 Howell Street
Code 814
Newport, RI 02841-1708
and

Theodore W. Frison
Randle, Inc.
P. O. Box 587
Great Falls, Virginia 22066

[1]Institute for Nonlinear Science

ABSTRACT

Data from experiments on the turbulent boundary layer around an axisymmetric vehicle rising under its own buoyancy are described in detail and analyzed using tools developed in nonlinear dynamics. Arguments are given that in this experiment the size of the wall mounted pressure sensors would make the data sensitive to the dynamics of about ten or so coherent structures in the turbulent boundary layer. Analysis of a substantial number of large, well sampled data sets indicates that the (integer) dimension of the embedding space required to capture the dynamics of the observed flows in the laminar regime is very large. This is consistent with there being no pressure fluctuations expected here and the signal being dominated by instrumental 'noise'. In a consistency check we find that data from the ambient state of the vehicle before buoyant rise occurs and data from an accelerometer mounted in the prow are also consistent with this large dimension. The time scales in those data are also unrelated to fluid dynamic phenomena.

In the **transition** and **turbulent** regions of the flow we find the pressure fluctuation time scales to be consistent with those of the fluid flow (about 240 to 250 μsec) and determine the dimension required for embedding the data to be about 7-8 for the transitional region and about 8-9 for the turbulent regime. These results are examined in detail using both global and local false nearest neighbor methods as well as mutual information aspects of the data. The results indicate that the pressure fluctuations are determined in these regimes by the coherent structures in the turbulent boundary layer. Applications and further investigations suggested by these results are discussed.

INTRODUCTION

Fluid flow over a rapidly moving body creates Tollmien-Schlichting (TS) waves which develop from the shear in the boundary layer near the body and grow by extracting energy from the mean flow into the complex behavior known generally as boundary layer turbulence. The turbulent drag on a body moving in this fluid is primarily due to this excitation of vortex motion, and the subject has properly been of substantial interest for many years. This paper reports on the analysis of data observed on a flow around an axisymmetric test vehicle while it rises under its own buoyancy. This configuration simplifies the flows by making them nearly two dimensional and still exposes flows of significant practical importance. In the study of open flows over flat plates the turbulent region is marked by coherent structures [1, 2] which appear partly as horseshoes of localized vorticity bent from lines of spanwise vortices by the mean flow, at least in the part of the boundary layer furthest from the wall, and appear as streamwise vortex structures near the wall. The appearance of these vortices and coherent structures even at high Reynolds number makes it plausible that only a few degrees of freedom might dominate the boundary layer flow even far downstream from the prow of the test vehicle. In the present case this idea is well established by the parameters of the flow and the observations about the coherent structures as summarized by Cantwell [1] and Robinson [2]. We return to this connection shortly.

The main result of the analysis in this paper is that methods which identify the nonlinear dynamical degrees of freedom in time domain reveal that in this experiment a small number of degrees of freedom are sensed by the pressure sensors flush mounted on the body to view pressure fluctuations in the boundary layer. While we shall have more to say about the boundary layer fluctuations, the fact that the sensors reveal only a few degrees of freedom to be active is consistent with having the many degrees of freedom (not residing in coherent or larger scale motions) significantly damped out in the observed regime of fluid flow. While small scale motions are present at some amplitude, they are "in the noise" relative to the observed degrees of freedom, and we should not expect to see them nor expect to have them play any significant role in the dominant dynamics of the phenomena seen by the sensors. This runs contrary to the idea that an enormous number of degrees are active in high Reynolds number flows. However, it is to be considered an aspect of the degrees of freedom that the sensors used in this experiment can actually sense with distinguishable amplitude. We shall argue that the mathematical idea that the dimension of the flow is enormous is not contradicted by the fact that real world sensors can be influenced by only a subset of those degrees of freedom, and this idea calls for the introduction of additional grounds for reasoning about the dimension of an observed time series which goes beyond the strict guidelines of the time delay embedding results [3, 4, 5] or the direct determination of dimension or active degrees of freedom from numerical simulations [6] which contain no consideration of the properties of the sensors involved in the observations.

DESCRIPTION OF THE EXPERIMENT

The experiment was conducted in a deep fresh water lake (Lake Pend Oreille) at a facility operated by the David Taylor Research Center/Acoustic Research Detachment in Bayview, Idaho. A buoyant test vehicle was used for the experiment. The test vehicle was propelled vertically from a depth of 1100 feet near the bottom of the lake by its own buoyancy. See Figure 1 for a representation of the experimental setup. The vehicle was an axisymmetric body 21 inches in diameter and approximately 27 feet in length. Included in the body was a weight section which held a symmetrically distributed array of cylindrical lead weights. Depending on the weight in this section, vehicle speeds of 40 to 75 feet per second (12.2 m/sec to 22.9 m/sec) were attained. In the experiment analyzed here the speed was 70 ft/sec (22.9 m/sec).

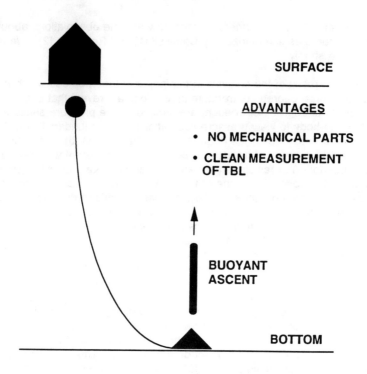

Figure 1. The set up of the experiment on the buoyant test vehicle. The vehicle was launched at ≈ 300 meters depth and rose freely to about 50 meters where it turned to avoid broaching the surface. Analog data was taken while the vehicle was in free flight, and sampled every 15.26 μsec for five seconds of data analysis.

The nose of the test vehicle (see Figure 2) was instrumented with piezoelectric pressure transducers arrayed along the axis of the test body to measure the pressure fluctuations associated with the developing boundary layer and the resulting turbulence. The transducers were PCB Model 112M149 having a sensitivity of 50 mV/psi (-26dB//1V/psi). The physical diameter of the transducers was 0.218 in (0.55 cm), and the effective diameter over which the instrument was sensitive to the flow, based on a rolloff of -6dB was found to be 0.12 in (0.30 cm) [7]. The pressure transducers were mounted in a carbon-graphite nose shell with stainless steel inserts as shown in Figure 3. The entire shell surface was covered with a 0.125 in (0.32 cm) elastomer in order to provide a smooth surface over which the boundary layer could develop. In this configuration the transducers did not trip the flow by interfering with the boundary layer.

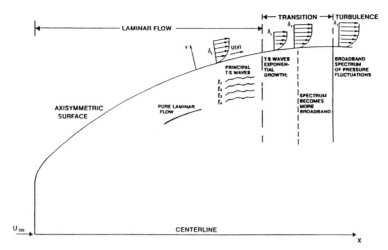

Figure 2. Schematic of the nose of the test vehicle showing the laminar, transition, and turbulent regions. Below the region designations is an indication of the physical processes dominant in that region. The vehicle was axisymmetric, so the designations hold around the body.

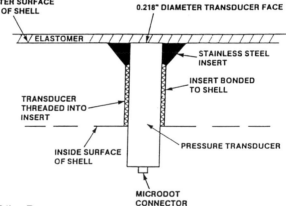

Figure 3. Configuration of the Pressure sensors.

The vehicle was hauled down to a depth of 1100 feet (335.5 meters) via cable attached to an onshore winch. Once the vehicle was stopped at the bottom of the lake, the onboard Honeywell 5600C tape recorder was powered on and ambient data from all sensors was recorded for 30 seconds. Following the ambient recording the vehicle released itself from the cable and ascended under its own buoyancy towards the lake surface. For a vehicle speed of 70 feet/sec, steady state conditions were achieved at a depth of 700 feet (213.5 meters). At 160 feet (48.8 meters) the vehicle rudders were activated in order to turn the vehicle horizontal so the surface would not be broached. The data reported on in this paper was taken during a five second interval of the steady part of the experimental run while the vehicle was between 550 and 200 feet below the surface.

The Fourier spectra of pressure fluctuations measured by transducers in laminar, transitional, and turbulent flow regimes are shown in Figures 4 to 8. Length, Reynolds number (Re_x), and displacement thickness, δ_* are also shown in the Figures. δ_* was computed using the Transitional Analysis Program System (TAPS)[(8)]. Because there are no measurable pressure fluctuations in laminar flow, the Fourier spectrum in that region consists of transducer response to the fluid loaded nose vibration and not to boundary layer pressure fluctuations. This will be clear in all of our later analysis as well, and it is important to keep this in mind as we proceed below. Transition and turbulent data are dominated by boundary layer fluctuations up to 3 or 4 kHz. Further discussion of the experimental situation can be found in [9].

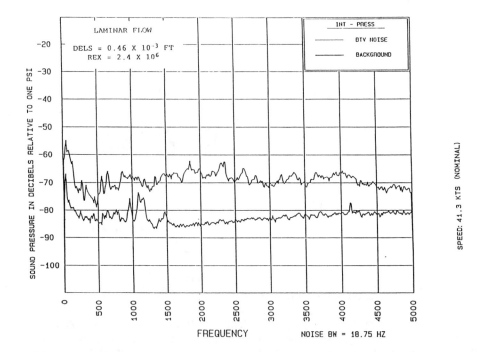

Figure 4. Fourier power spectrum of the time series for sensor B1 located in the laminar zone of the flow. Spectra are shown for the ambient pressure when the vehicle was at rest (this is the heavy solid line designated as BACKGROUND) and for flow conditions as described in the text (light solid line).

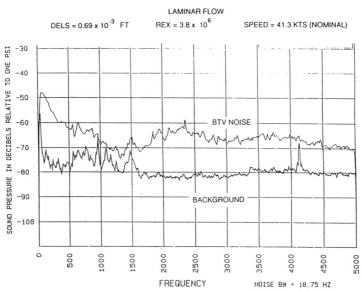

Figure 5. Fourier power spectrum of the time series for sensor B3 located in the laminar zone of the flow. Spectra are shown for the ambient pressure when the vehicle was at rest (this is the heavy solid line designated as BACKGROUND) and for flow conditions as described in the text (light solid line).

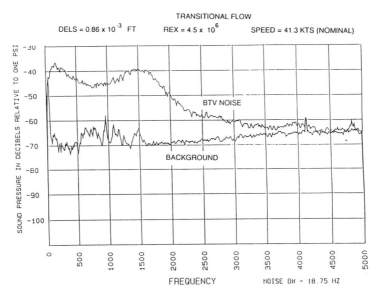

Figure 6. Fourier power spectrum of the time series for sensor B4 located in the transition zone of the flow. Spectra are shown for the ambient pressure when the vehicle was at rest (this is the heavy solid line designated as BACKGROUND) and for flow conditions as described in the test (light solid line).

62 Nonlinear Analysis of High Reynolds Number Flows

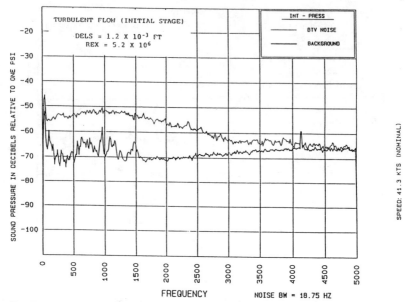

Figure 7. Fourier power spectrum of the time series for sensor B5 located in the turbulent zone of the flow. Spectra are shown for the ambient pressure when the vehicle was at rest (this is the heavy solid line designated as BACKGROUND) and for flow conditions as described in the text (light solid line).

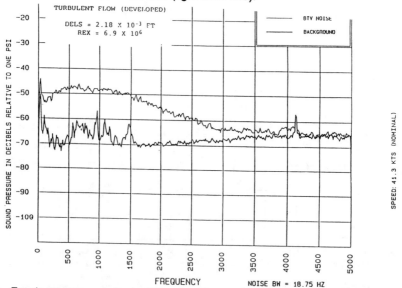

Figure 8. Fourier power spectrum of the time series for sensor B7 located in the turbulent zone of the flow. Spectra are shown for the ambient pressure when the vehicle was at rest (this is the heavy solid line designated as BACKGROUND) and for flow conditions as described in the text (light solid line).

The turbulent boundary layer thickness ranged from .036 to 0.85 cm through the region of the pressure sensors. Using a nominal speed of the vehicle of u ≈ 21 meters/sec, (or convective velocity, u_c, of approximately 15 meters/sec) we deduce a typical turnover time or inverse shear in the boundary layer of δ/u or $δ/u_c$ ≈ 240 - 250 μsec. The turnover time is computed at a spatial point at the onset of turbulence. The importance of the outer flow variables in the behavior of the pressure fluctuations seen on the wall in boundary layer flows is emphasized by the results of Farabee and Casarella [10]. Also, the scaling of vorticity variations, clearly the dominant feature in the boundary layer coherent structures, is much cleaner when done with outer variables [11]. The Reynolds number, based on boundary layer thickness, uδ/ν, and these other dimensions is about 10^5. The sampling time of $τ_s$ =15.26 μsec chosen for the acquisition of the data guarantees we will capture the main variations in the pressure due to fluid motions, and the shear rate of ≈ 4 kHz means we will lose little by the low pass filtering done at 6400 Hz.

TOOLS USED IN ANALYSIS OF THE EXPERIMENT

We have performed analysis of numerous data sets from these experiments using four to five seconds of data in each set. This results in 262,144 points for four seconds of data and 327,680 points for five seconds of data. The tools we have used to examine the data include the familiar examination of the time traces and the Fourier spectra of these. We will show that these tools reveal useful but limited information about the dynamics. We have also studied characteristics of the nonlinear motion [5, 12] with:

• **average mutual information** which determines the way in which measurements at time t nonlinearly correlate [13] with measurements at time t´≠t. We will show that the typical information decorrelation time in the **transition and turbulent** regions is about 15 $τ_s$ ≈230 μsec. This is consistent with the fluid dynamical time scales associated with the inverse shear estimated above,

• **global false nearest neighbors** which determines in what dimension the attractor in this fluid flow is unfolded [14] in a coordinate system composed of time lags of the pressure measurements taken at lags determined by the average mutual information,

• **local false nearest neighbors** which tells us the dimensions of the dynamical model to be used to describe the data [15].

We began the evaluation of the local and global Lyapunov exponents for these attractors [5], but even with the very large data sets we did not convince ourselves that we had sufficient data for the accurate evaluation of these important quantities. We are continuing work on these as they dictate the predictability of the data we are observing. The issue has to do with the evaluation of these exponents in dimensions as high as eight to ten. We did conclude that in the transition region where the dimension is lower that at least one exponent is positive, consistent with the idea of low dimensional chaos, and has a value about $(10\ τ_s)^{-1}$. One exponent is always zero which tells us that the dynamics are governed by a set of differential equations. We shall return to these important questions in a subsequent article.

SCALES IN THE EXPERIMENT

In the Cantwell and Robinson review articles [1, 2] we find the introduction of 'wall units' to be an enormously useful way to capture qualitatively the scales of any given experiment. These units are defined in terms of the stress τ_w at the wall which in turn defines a friction velocity u_τ

$$u_\tau = \sqrt{(\tau_w/\rho)} \qquad (1)$$

where ρ is the density of the fluid. From this velocity and the kinematic viscosity ν of the fluid, a length scale

$$\nu/u_\tau \qquad (2)$$

is introduced. In the present experiment we establish a feel for the size of things by using the values for these quantities at the location of the sensor B7 (location to be given below) where ambient data as well as pressure fluctuations during buoyant rise of the body was made available for analysis. Using the value $\nu = 1.57 \times 10^{-2}$ cm^2/sec for water and the value $u_\tau = 72.8$ cm/sec deduced from the experimental setup and TAPS [8], we find a length scale for 'wall units' of 2.16×10^{-4} cm. In these units the effective sensor size is about 1400 wall units.

The distance between streamwise coherent structures in the turbulent boundary layer has been observed to be about 100 wall units. Similarly the size of a horseshoe or hairpin structure is around 150 to 200 wall units. With these estimates we see that on the order of ten coherent structures would be within the sensitivity area of the sensors used in the experiment. On these grounds we would expect that the dynamics of these few coherent structures should totally dominate the observations here. This is further supported by the observations of Schewe [16] who analyzed the ability of pressure sensors on the wall to identify all degrees of freedom in a turbulent boundary layer flow. He concluded that when the sensor is about 20 wall units or less it picks up the finest details of the flow, but as one increases the sensor size it becomes sensitive to other aspects of the flow. Schewe studied sensor sizes from 19 to 333 wall units in diameter. Our sensors are 70 times bigger than this suggested size of 20 wall units, and it is quite plausible that they average out pressure fluctuations on the smallest scales and report only effects due to the larger scale coherent structures of which we estimate order of ten are active in the sensors range.

Finally, our Fourier spectra for the transition and turbulent regions in the vicinity of frequencies given by $2 \pi f \delta / u_\tau \approx 50$ to 70 which is just in the region that Farabee and Casarella [10] identify as the signature of the maximum effect of outer flow dynamics on the wall pressure fluctuations. Defining the outer scale frequency by $f_0^{-1} = 2 \pi f \delta / u_\tau$, we have $f_0 \approx 23.2$ Hz. The region $f \leq 5 f_0 \approx 116$ Hz is called the low frequency regime in [10] and identified as coming from large distances from the wall and contributing less than 1% of the total RMS pressure fluctuations. These frequencies were filtered out of the present data in any case. The region $f \approx 50\ f_0$ is called the mid-frequency region in [10], and their analysis quite clearly associates this regime with turbulent activity in the outer region of the boundary layer. This is consistent with the other scale estimates we have made here.

DATA ANALYSIS

We will present the analysis of many data sets from the experiments just described. The first group are the stations designated as B1, B3, B4, B5 and B7 along one of the axial lines of sensors arrayed on the buoyant test body. We also look at data from the sensor at B7 when the test vehicle was sitting motionless at the lake bottom and data from the fore mounted accelerometer during motion. The ambient data sets a scale and qualitative description of the 'noise' in the system. The accelerometer gives a look at the excitation of the structural modes of the body, and this will be seen in contrast to the fluid fluctuations observed by the pressure sensors. The data sets B1 and B3 are in the 'laminar' regime where pressure fluctuates from the fluid dynamical pressure fluctuations in the laminar region, so what we see here are either instrumental fluctuations or pressure associated with fluid-loaded vibrations of the hull of the buoyant body. The data set B4 is in the 'transition' regime, and the sets B5 and B7 are in the 'turbulent' regime. To assure ourselves that conclusions about one region of flow or another are not specific to the line of 'B' sensors, we have also looked at data sets from stations C2, F3 in the 'laminar' regime, stations C4, C5 and E4 in the 'transition' regime and A8 and A7 in the 'turbulent' regime. We will not display the complete analyses for these stations as the results do not further illustrate the points to be made with the B-station data. Since the results for the transition and turbulent regions are quite striking, we will present material from the analysis of the sensors at stations C5 and A8. The former is in the transition region and the latter near the end of the sensor chain in the turbulent region. The behavior of the pressure fluctuations from these and all the other stations is entirely consistent with that presented from the B-stations in the respective regimes.

Chaos is a phenomenon in multivariate state space. It may be observed as a scalar time series as we do here in output voltages from a sensor, but we need to reconstruct a many dimensional phase space for viewing the chaotic structure and for computing distinguishing characteristics of the data. The method we use for this is called time delay state space reconstruction [3, 4, 5, 12] and consists of making d-dimensional vectors from the time delays of the observations. Thus from measured voltages $v(n) = v(t_0 + n\tau_s)$ we make vectors

$$y(n) = [v(n), v(n+T), v(n+2T), \ldots v(n+(d-1)T)] \qquad (3)$$

and the first task is to find T and d.

AMBIENT DATA

We begin with the data set taken at station B7 before the vehicle was launched. This is the ambient data. In Figure 9 we show the time trace of the voltage measured as the output of the pressure transducer as a function of time. This is about 10% of the recorded data stream. In Figure 10 we have the Fourier power spectrum for this data. The frequency axis has units of 8 Hz, so the roll off of the spectrum at frequency 800 is just the effect of the 6400 Hz low pass filtering.

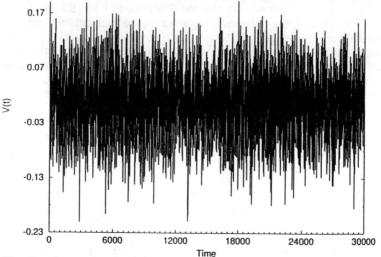

Figure 9. The time trace of voltage from the pressure transducer at location B7 in the ambient state. The vehicle was at rest.

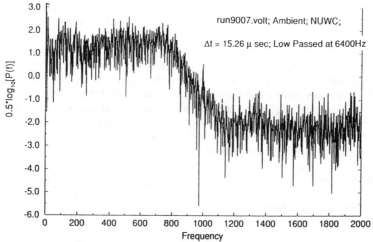

Figure 10. The Fourier power spectrum from voltage measurements at the pressure transducer at location B7 in the ambient state. The vehicle was at rest.

AVERAGE MUTUAL INFORMATION

Next we will display the average mutual information calculation for the ambient data. At this juncture we will digress to explain the statistic we are computing so the reader can understand what we are evaluating.

Nonlinear systems in a parameter regime where the orbits are chaotic are known to generate entropy in the direct sense of Shannon (5). This suggested to several authors [13] that the information theoretic properties of chaotic systems would be particularly useful in their study. The average mutual information is one of those tools. It answers the question: if we have made a measurement of voltage from the pressure sensors $v(n) = v(t_0+n\tau_s)$, how much information (in bits) do we have about the measurement of voltage (pressure) a time $T\tau_s$ later; namely, $v(n + T)$. The information theoretic answer to this question requires the distribution of the measurements $v(n)$ and $v(n + T)$ (the distribution of these is the same) over the set of measured data and the joint distribution of measurements of these two quantities. The first we call P $(v(n))$, the second, P $(v(n+T))$, and the last, $P(v(n), v(n+T))$. The mutual information between these measurements is

$$\log_2[P(v(n),v(n+T)) / P(v(n))P(v(n+T))] \qquad (4)$$

and the average over all measurements is

$$I(T) = \sum_{n=1}^{N} P(v(n), v(n+T)) \log_2[P(v(n), v(n+T))/P(v(n))P(v(n+T))] \qquad (5)$$

when we have N observations. I(T) is the average mutual information. If the measurements $v(n)$ and $v(n + T)$ are independent, then each term in this sum vanishes since the joint probability factorizes $P(a,b) = P(a) P(b)$. We expect the two measurements to become independent for very large T since chaotic signals rapidly lose memory of earlier entries on their orbits. $I(T = 0)$ is large, reflecting the full knowledge we have of the measurements. The actual value, I(0), is the Shannon entropy. In general $I(T) > 0$ and we seek some intermediate value of T where I(T) is not too large or too small. If we can find such a value, then that T will serve as a candidate for determining rather independent measurements $v(n)$ and $v(n + T)$ - independent in a nonlinear sense. A nonlinear prescription, similar in spirit to the prescription often used in linear analysis of choosing the first zero of an autocorrelation function, is to choose the first minimum of I(T). In practice any time lag in the vicinity of this minimum will do fine, and the mathematical theorem which underlies this construction [3, 4] is true, in principle, independent of T.

Evaluating I(T) gives the data shown in Figure 11. This function has its first minimum at T = 5 or at a lag of approximately 76 μsec. The region of this minimum is shown in expanded format in Figure 12. From our earlier comments we would conclude that this delay has little, if anything, to do with time scales associated with the fluid flow.

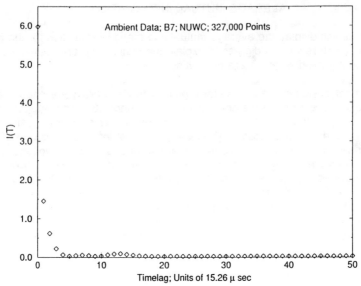

Figure 11. The average mutual information evaluated from data taken at station B7 in the ambient state. The first minimum of I(T) is at T = 5 or 76 μsec.

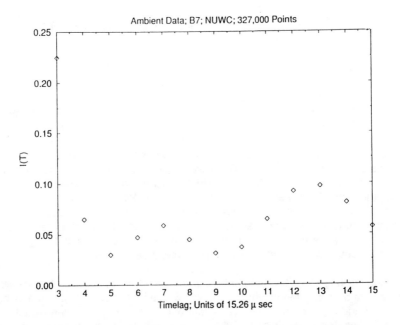

Figure 12. An expanded scale view of the average mutual information evaluated from data taken at station B7 in the ambient state. The first minimum of I(T) is at T = 5 or 76 μsec.

FALSE NEAREST NEIGHBORS

To determine an appropriate value for the dimension of the state space in which we will view the observed process, we use a method which inquires into the geometric basis for the theorem of Mane and Takens [3, 4]. The idea is that we observe the data from a multivariate structure projected down onto the observation axis: v(n) here. To unfold this structure we must add additional coordinates for the space. We have added enough additional coordinates when all points are near each other for dynamical reasons rather than because they got there by projection from a higher dimension. We proceed [14] by determining in dimension 'd 'which points made out of time delays into vectors as above are the nearest neighbors $y^{NN}(n)$

$$y^{NN}(n) = [v^{NN}(n), v^{NN}(n+T), ..., v^{NN}(n+(d-1)T)] \qquad (6)$$

of the point y(n). Then we ask whether these points remain near in dimension (d + 1) where the vector y(n) is augmented by a component v(n +dT) and $y^{NN}(n)$ is augmented by $v^{NN}(n + dT)$. If this distance is small, then the neighbors are true neighbors. If not, then we have false neighbors which arrived near each other by projection. When the percentage of false nearest neighbors falls to zero, we have unfolded the attractor.

'Noise', which we have come to understand is high dimensional dynamics, will have a large percentage of false nearest neighbors for any low dimension, say up to 20 or so where we typically stop computing. Eventually, as the number of dimensions reaches that of the dynamical rule generating the 'random' numbers, the percentage of false nearest neighbors will drop to zero. If we add a high dimensional signal to a low dimensional signal [5, 14], then the false neighbors will fall for a while and eventually rise. These qualitative statements are dependent on the number of data, but seem to hold as a descriptive feature of this kind of data.

In Figure 13, we show the percentage of false nearest neighbors for the ambient data using the time delay of T = 5 indicated above and using first 17,000 and then 87,000 data points. It is clear from this graph that the ambient data represent a high dimensional signal. Indeed, from this test, with the computing power we presently have, and the number of data available, we cannot say in what dimension the data may be unfolded, except to agree that it is higher than 20. In Figure 14 we show for comparison the false nearest neighbor computation for the Lorenz model [17] which is composed of three ordinary differential equations. The computation is shown for the clean chaotic signal, the same signal with 50% (S/N = 6 dB) uniform random noise added, and the same signal with 100% (S/N = 0 dB) uniform random noise added.

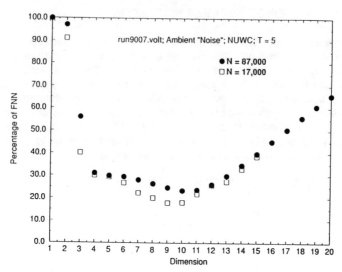

Figure 13. Global false nearest neighbors for the data from station B7 when the vehicle was at rest. The time delay of T = 5 used in constructing phase space vectors y(n) is taken from the first minimum of the average mutual information. Results are shown for N = 17,000 and N = 87,000 data points. The implication that the ambient data is high dimensional is independent of the number of data in this range.

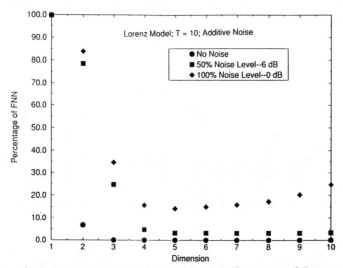

Figure 14. Global false nearest neighbors for data from one of the components of the three degree of freedom Lorenz model [17]. The false nearest neighbors are shown when the data is clean, when we have added uniform random numbers with an RMS level of 50% of the size of the attractor, and when we have added noise at 100% the size of the attractor. The false nearest neighbor calculation degrades gracefully with contamination. This figure suggests that in some of the data analyzed in this paper we may be seeing a low dimensional system contaminated by high dimensional 'noise'.

It is clear from this example, that the ambient signal may be composed of a low dimensional signal plus some high dimensional 'noise', but no quantitative test for this kind of conclusion is yet available. The qualitative suggestion, however, arguing by analogy with the behavior of the Lorenz data contaminated at various levels is both suggestive and interesting.

The false nearest neighbor test thus establishes here that this ambient data is 'noise-like, and we can proceed no further with our analysis of this data using the tools presently available for nonlinear systems. More precisely, when a signal is shown to be high dimensional, the analysis tools for working further with that signal are basically not well developed at this time. The distinction between low and high dimensional is not one made in principle but in practice and the qualitative break point is about dimension eight to ten. In any case, we will see quite a different behavior for transition and laminar data. We have presented this in detail since when we come to other of the test vehicle data sets, the contrast will be striking.

'LAMINAR' DATA

Next, we examine data from the 'laminar' region. The time series of voltage from station B1 is shown in Figure 15. It is irregular, appearing much as the ambient data. Note the low level of voltage (the proxy for pressure) fluctuations in this region. This is consistent with the designation laminar, where no pressure fluctuations should occur. In Figure 16, we have the Fourier power spectrum of this data. The rolloff above 6400 Hz is evident, and the spectrum is otherwise broad and uninformative. In Figure 17, we have the average mutual information for this data set; zero lag is suppressed. We can clearly see a minimum at T = 7, or 7 $\tau_s \approx$ 107 μsec. This is similar to the first minimum for the ambient data set, and indicates that the time delay here is characteristic of something other than the fluid flow. Next we examine the false nearest neighbors for this data. In Figure 18, we show this for dimensions 1< d < 20 using 87,000 data points from the measurements. Once again we cannot say what dimension should be used for unfolding an attractor for this data, except that it is quite high. One might conclude that there is some evidence that an underlying dynamical process of some low dimension, of order seven or so, has been seen here contaminated by substantial amounts of 'noise from other sources. False nearest neighbors is not a fine tuned enough tool to make that kind of conclusion firm.

72 Nonlinear Analysis of High Reynolds Number Flows

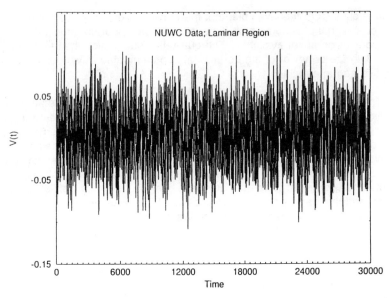

Figure 15. Time series of voltage from the pressure transducer located at station B1 in the laminar region during the rise of the vehicle. Note the low voltage levels compared to the transition and turbulent regions seen below. In noise free fluid flow there should be no pressure fluctuations in the laminar region.

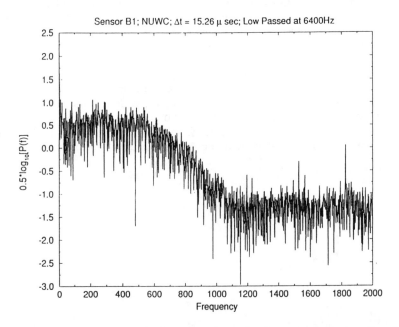

Figure 16. The Fourier power spectrum from the voltages measured at station B1 during the rise of the vehicle. The frequency axis is in units of 8 Hz, and the rolloff at 6400 Hz is caused by the filtering of the data during the experiment.

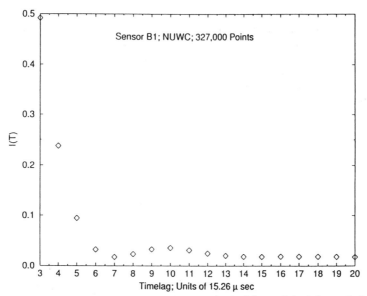

Figure 17. The average mutual information evaluated from data taken at station B1 in the laminar zone during the rise of the vehicle. The first minimum of I(T) is at T = 7 or 107 μsec.

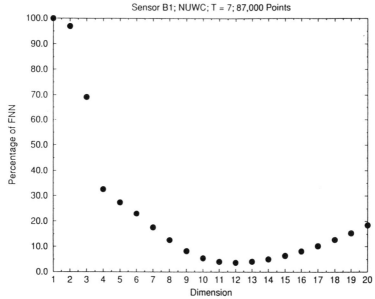

Figure 18. Global false nearest neighbors for the data from station B1 when the vehicle was in motion. The time delay of T = 7 used in constructing phase space vectors y(n) is taken from the first minimum of the average mutual information. Results are shown for N = 87,000 data points.

A similar set of results applies for the data from station B3 which is also in the laminar region. Neither the time series nor the power spectrum is very revealing. In Figure 19 are the results for average mutual information for this data station. There is again a clear minimum at T = 7, and that is the one we use in reconstructing the vectors in multivariate phase space. In Figure 20 is the result for false nearest neighbors for this data. We reach the same conclusions just as before for the other laminar flow data.

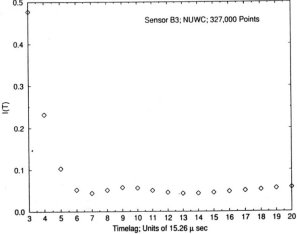

Figure 19. The average mutual information evaluated from data taken at station B3 in the laminar zone during the rise of the vehicle. The first minimum of I(T) is at T = 7 or 107 μsec.

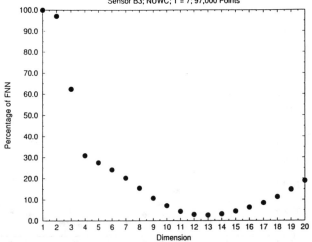

Figure 20. Global false nearest neighbors for the data from station B3 when the vehicle was in motion. The time delay of T = 7 used in constructing phase space vectors y(n) is taken from the first minimum of the average mutual information. Results are shown for N = 97,000 data points.

In a sense this is surprising, and in another sense this is natural. Of course, one expects laminar flow to be regular and low dimensional. Indeed, it is. From the point of view of pressure fluctuations perfectly clean data should have dimension zero; that is, no pressure fluctuations in laminar flow. However, we have data dominated by the transducer response to fluid-loaded nose vibrations. The indications of this are the levels of the fluctuations, and even more telling is the place where average mutual information has its first minimum (namely $T \approx 5$-7) characteristic of some time scale other than that of the fluid flow. Finally, the false nearest neighbor test shows that each of the ambient and laminar data sets is very high dimensional.

"TRANSITION" REGION DATA

Now we come to much more interesting data. In the transition region we should see some evidence of the production of T-S waves, and see some significant pressure fluctuations. The data is now from station B4, and the time series is displayed in Figure 21. Figure 22 is the Fourier power spectrum for this data. In the latter we see a signal around a frequency of 1500 Hz which can be attributed to the excitation of T-S waves in this system. The spectrum associated with the T-S excitations is broad, indicating substantial interaction among the T-S modes. In Figure 23 we display the average mutual information which has its first minimum at $T = 16$ or about 244 μsec which is now characteristic of the fluid flow according to our earlier estimates. In Figure 24 we show the plot of false nearest neighbors for this data for $1 < d < 20$. This is substantially different from the behavior of earlier data sets. Here the percentage of false nearest neighbors drops to zero at $d = 7$ and then remains there. This behavior is characteristic of a low dimensional chaotic system whose attractor has been unfolded at dimension 7.

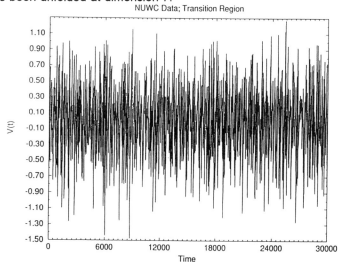

Figure 21. Time series of voltage from the pressure transducer located at station B4 in the transition region during the rise of the vehicle. Note the high voltage levels compared to the laminar regions seen above.

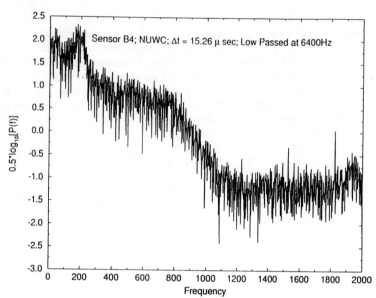

Figure 22. The Fourier power spectrum from the voltages measured at station B4 in the transition zone during the rise of the vehicle. The frequency axis is in units of 8 Hz, and the rolloff at 6400 Hz is caused by the filtering of the data during the experiment. The broad peak in the neighborhood of 1500 Hz is attributed to T-S wave excitation.

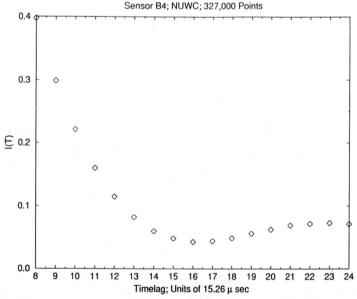

Figure 23. The average mutual information evaluated from data taken at station B4 in the transition zone during the rise of the vehicle. The first minimum of I(T) is at T = 16 or 244 μsec. This is a time scale consistent with the fluid dynamics of the boundary layer.

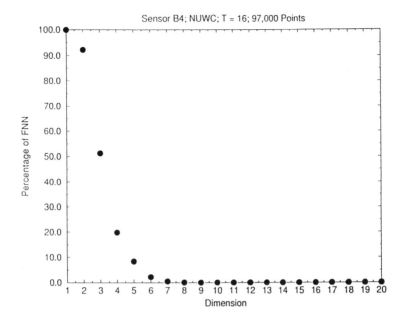

Figure 24. Global false nearest neighbors for the data from station B4 the vehicle was in motion. The time delay of T = 16 used in constructing phase space vectors y(n) is taken from the first minimum of the average mutual information. Results are shown for N = 97,000 data points. The fall of false nearest neighbors to zero at dimension seven is a result of low dimensional dynamics as the source of the pressure fluctuations.

Since this is so different from the previous data sets, even though the time series and power spectra are not that different, we take a look at another data set from the transition region. The set we chose was from station C5 which is located very near the end of the transition region. In the power spectrum for this sensor the region around 1500 Hz where TS waves were located before is less distinct, but then we are further down the body than at B4 so we have entered a region of more pronounced nonlinear interaction among the modes. The average mutual information for the C5 data set is shown for lags 12 < T < 22 in Figure 25 where the minimum, the first in this data, is seen at T = 16. This is the same as in the previous transition data set. In Figure 26 the false nearest neighbors for this data set is shown for 1 < d < 12 and in Figure 27 this is blown up for 5 < d < 12. The percentage of false nearest neighbors has dropped below 0.5% at d = 7 indicating we almost certainly have unfolded the attractor, though the cautious person would wish to choose d = 8. In a moment we will provide further evidence that d = 8 is appropriate. In any case, we see a clear statement that the number of dimensions required to capture this data set, as in the case of station B4, is small, namely about seven or eight. This is in sharp contrast to the laminar or ambient data. Further underlining this as a dynamical feature is the distinguishably higher level of pressure fluctuations seen in the time series and the fluid dynamical relevance of the time delay T = 16 in units of τ_s.

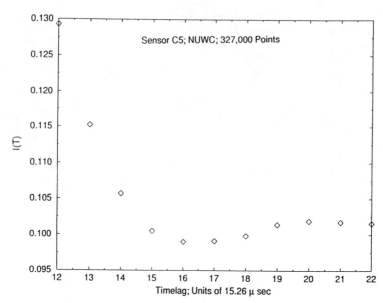

Figure 25. The average mutual information evaluated from data taken at stations C5 in the transition zone during the rise of the vehicle. The first minimum of I(T) is at T = 16 or 244 μsec. This is a time scale consistent with the fluid dynamics of the boundary layer.

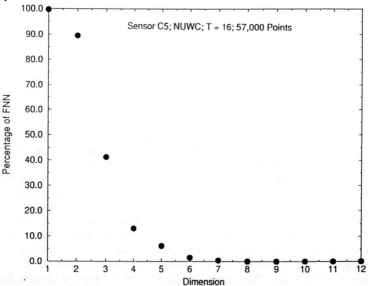

Figure 26. Global false nearest neighbors for the data from station C5 when the vehicle was in motion. The time delay of T = 16 used in constructing phase space vectors y(n) is taken from the first minimum of the average mutual information. Results are shown for N = 57,000 data points. The fall of false nearest neighbors to zero at dimension seven is a result of low dimensional dynamics as the source of the pressure fluctuations.

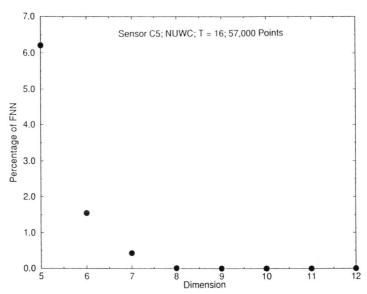

Figure 27. Expanded scale for global false nearest neighbors for the data from station C5 when the vehicle was in motion. 5<d<12. The time delay of T = 16 used in constructing phase space vectors y(n) is taken from the first minimum of the average mutual information. Results are shown for N = 57,000 data points.

LOCAL FALSE NEAREST NEIGHBORS

As another tool for examining the data here we look at a quantity which asks how many dynamical degrees of freedom are excited locally on the attractor. The global false nearest neighbor criterion produces a global number which allows the unfolding of the attractor in the time delay coordinate system. However, the time delay coordinates are almost certainly not the original coordinates in which the system evolves and it is quite plausible that the embedding dimension associated with the global false nearest neighbors is larger than that of the dynamics itself. To examine this, we have created a local false nearest neighbor test [15] which examines in every neighborhood on the attractor how well one can predict ahead the evolution of two neighboring points in dimensions less than or equal to the dimension given by the global false nearest neighbor test. If the percentage of bad predictions which may be due to numerical accuracy or the quality of the particular prediction method used becomes independent of local dimension, then that dimension is picked out as the dimension of the dynamics. The basic idea is that when neighbors are false they are nearby for geometric, not dynamical reasons, so they will lead to bad predictions because they will evolve rapidly to far separated parts of the attractor. True neighbors will move along with each other and the quality of one's ability to predict where they will go is not limited by their having been projected together from a higher dimension where they are, in fact, quite well separated. In Figure 28 we show the percentage of bad predictions for data set C5 as a function of local dimension and of number of neighbors. This shows that at d = 8, which is the same dimension as the global false neighbor test indicated, that the predictability becomes independent of these variables, thus indicating that dimension 8 is correct for this set of observations.

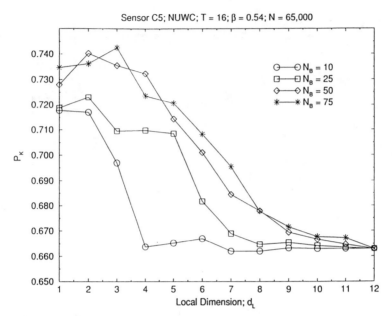

Figure 28. Local false nearest neighbors for the data from station C5 in the transition zone. The time delay T=16 comes from average mutual information. N=75,000 data points were used. β measures the fraction of the attractor over which a bad prediction must occur in the local false nearest neighbor calculation. The percentage of bad predictions becomes independent of the number of neighbors N_B and the local dimension near $d_L=9$. $N_B=10, 25, 50$ and 75 were used here.

'TURBULENT' DATA SETS

Now we examine data from two turbulent data sets. The first is station B5 which is located at the beginning of the turbulent region. In Figure 29, we have the power spectrum at this sensor. In Figure 30 the average mutual information is shown from $5 \leq T \leq 22$ and a clear minimum at T = 16 is revealed. T = 16 is a time of 244 μsec which is consistent with fluid flow dynamics. The amplitude of the pressure fluctuations, compared to the ambient or laminar data, supports this conclusion. In Figure 31 we have the global false nearest neighbors result for station B5. It is clear that at dimension nine or ten we have unfolded the attractor completely. We further examine this conclusion by a look in Figure 32 at the local false nearest neighbors for data set B5. Here it is clear that at local dimension nine we have achieved the independence of local dimension and number of neighbors which is characteristic of having unfolded the attractor locally.

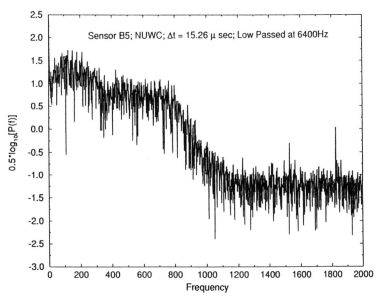

Figure 29. The Fourier power spectrum from the voltages measured at station B5 in the turbulent zone during the rise of the vehicle. The frequency axis is in units of 8 Hz, and the rolloff at 6400 Hz is caused by the filtering of the data during the experiment.

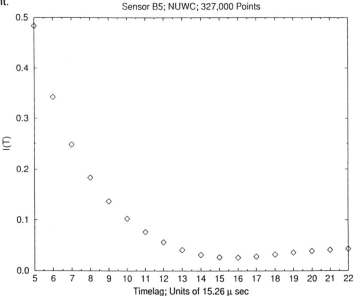

Figure 30. The average mutual information evaluated from data taken at station B5 in the turbulent zone during the rise of the vehicle. The first minimum of I(T) is at T = 16 or 244 μsec. This is a time scale consistent with the fluid dynamics of the boundary layer.

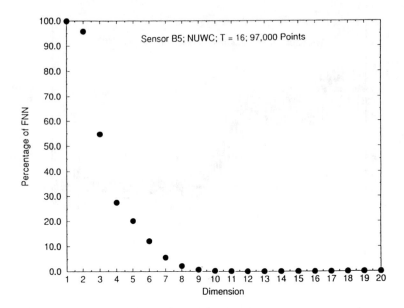

Figure 31. Global false nearest neighbors for the data from station B5 when the vehicle was in motion. The time delay of T = 16 used in constructing phase space vectors y(n) is taken from the first minimum of the average mutual information. Results are shown for N = 97,000 data points. The fall of false nearest neighbors to zero at dimension nine is a result of low dimensional dynamics as the source of the pressure fluctuations.

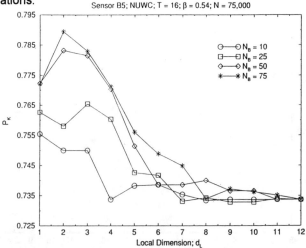

Figure 32. Local false nearest neighbors for data from station B5 in the turbulent zone. The time delay T = 16 comes from average mutual information. N = 75,000 data points were used. β measures the fraction of the attractor over which a bad prediction must occur in the local false nearest neighbor calculation. The percentage of bad predictions becomes independent of the number of neighbors N_B and the local dimension near d_L = 9, N_B = 10,25,50 and 75 were used here.

Our next example is data from station B7 located in fully developed turbulent flow. There are no surprises in the time series or the Fourier power spectrum. Figure 33 shows the average mutual information from $10 \leq T \leq 25$, and the minimum at $T = 18$ (the first minimum) is evident. Using this value of the timelag we evaluate the global false nearest neighbors which is shown in Figure 34; a global embedding dimension of $d = 8$ is revealed for this data. In Figure 35 we display the local false nearest neighbor calculation for data set B7. This makes it quite clear that at dimension eight we have locally unfolded the attractor and confirms the evidence from the global examination of this question.

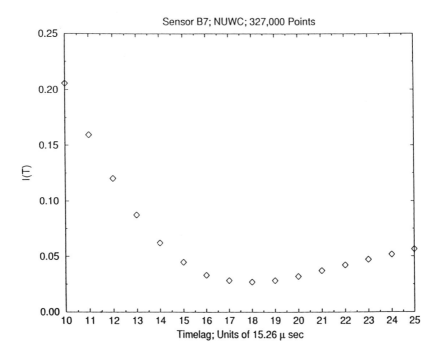

Figure 33. The average mutual information evaluated from data taken at station B7 in the turbulent zone during the rise of the vehicle. The first minimum of I(T) is at $T = 18$ or 275 μsec. This is a time scale consistent with the fluid dynamics of the boundary layer.

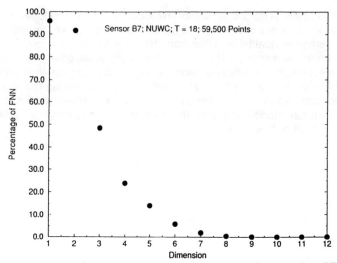

Figure 34. Global false nearest neighbors for the data from station B7 when the vehicle was in motion. The time delay of T = 18 used in constructing phase space vectors y(n) is taken from the first minimum of the average mutual information. Results are shown for N = 59,500 data points. The fall of false nearest neighbors to zero at dimension eight is a result of low dimensional dynamics as the source of the pressure fluctuations.

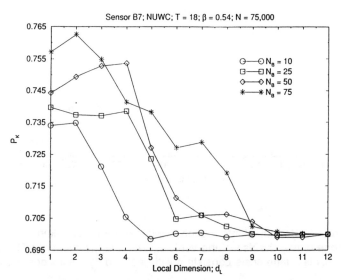

Figure 35. Local false nearest neighbors for data from station B7 in the turbulent zone. The time delay T = 16 comes from average mutual information. N = 75,000 data points were used. β measures the fraction of the attractor over which a bad prediction must occur in the local false nearest neighbor calculation. The percentage of bad predictions becomes independent of the number of neighbors N_B and the local dimension near d_L = 9. N_B = 10, 25, 50 and 75 were used here.

The last look at data from pressure sensors comes from examining data from station A8 which is within the fully turbulent region. In Figure 36 we show the average mutual information from this sensor. This reveals a clear minimum at T = 20 which corresponds to a timelag of 300.5 μsec, which is associated with the fluid flow. Using this value of T, we determine the percentage of global false nearest neighbors which is shown in Figure 37. The percentage drops very near zero by dimension 7, and we zoom in on this data in Figure 38 where global false nearest neighbors is shown for $6 \leq d_e \leq 15$ and we see the percentage of false neighbors go to zero at d = 9 and then stay there. The local false nearest neighbors test displayed in Figure 39 confirms that at dimension nine we have removed all ambiguities in true neighbors by unfolding the attractor.

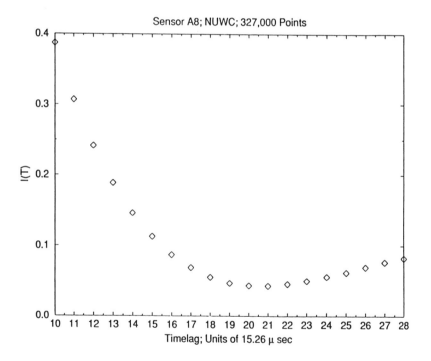

Figure 36. The average mutual information evaluated from data taken at station A8 in the turbulent zone during the rise of the vehicle. The first minimum of I(T) is at T = 20 or 305 μsec. This is a time scale consistent with the fluid dynamics of the boundary layer.

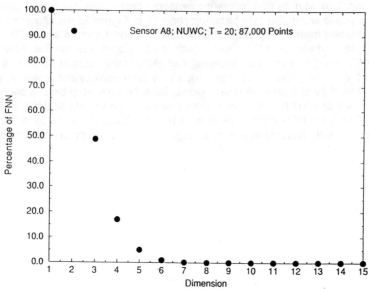

Figure 37. Global false nearest neighbors for the data from station A8 when the vehicle was in motion. The time delay of T = 20 used in constructing phase space vectors y(n) is taken from the first minimum of the average mutual information. Results are shown for N = 87,000 data points. The fall of false nearest neighbors to zero at dimension seven is a result of low dimensional dynamics as the source of the pressure fluctuations.

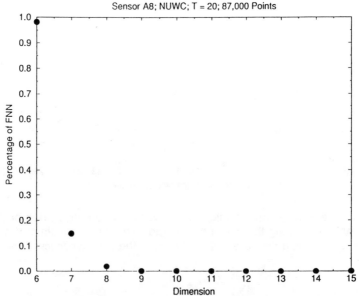

Figure 38. Expanded scale for global false nearest neighbors for the data from station A8 when the vehicle was in motion. $6 \leq d \leq 15$. The time delay of T = 20 used in constructing phase space vectors y(n) is taken from the first minimum of the average mutual information. Results are shown for N = 87,000 data points.

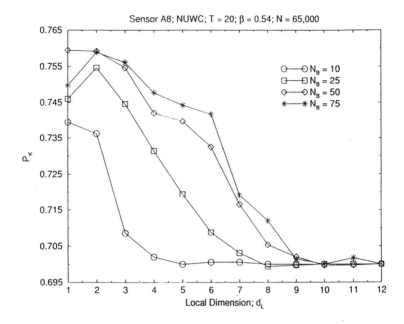

Figure 39. Local false nearest neighbors for data from station A8 in the turbulent zone. The time delay T = 20 comes from average mutual information. N = 65,000 data points were used. β measures the fraction of the attractor over which a bad prediction must occur in the local false nearest neighbor calculations. The percentage of bad predictions becomes independent of the number of neighbors NB and the local dimension near dL = 9. NB = 10,25,50 and 75 were used here.

ACCELEROMETER DATA

We have one other glimpse of the dynamics in this experiment, namely data from an accelerometer mounted forward in the test vehicle beneath a region of turbulent flow. In Figure 40 we have the output voltage from the sensor as a function of time, while in Figure 41 we have the Fourier power spectrum for this data. Again this is uninformative about the dynamics; the rolloff at 6400 Hz from the low pass filter is evident. In Figure 42 for $3 \leq T \leq 20$ we see the first minimum of the average mutual information at T = 5 which does not correspond to fluid dynamical time scales. Finally using this data, we show in Figure 43 the global false nearest neighbors which demonstrates that this is very high dimensional dynamics and significantly different from the dynamics of the fluid flow seen in the transitional turbulent data sets. The accelerometer data is consistent with the laminar data which represents the response of the transducer to nose vibration.

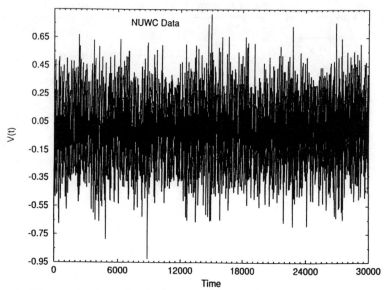

Figure 40. Time series from the voltages on the accelerometer mounted in the prow of the buoyant test vehicle while the vehicle was in motion.

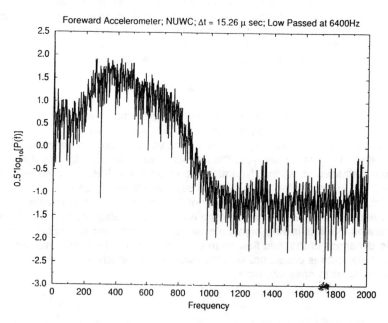

Figure 41. The Fourier power spectrum from the voltages measured at the forward mounted accelerometer during the rise of the vehicle. The frequency axis is in units of 8 Hz, and the rolloff at 6400 Hz is caused by the filtering of the data during the experiment.

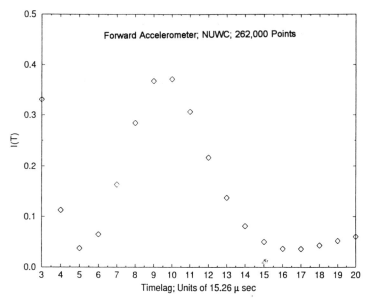

Figure 42. The average mutual information evaluated from data taken at the forward mounted accelerometer during vehicle motion. The first minimum of I(T) is at T = 5 or 76 μsec.

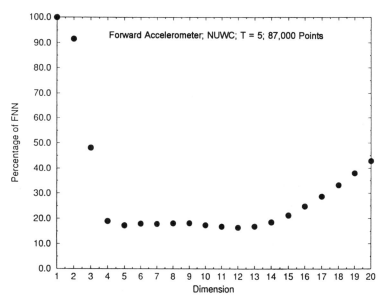

Figure 43. Global false nearest neighbors for the data from the forward mounted accelerometer when the vehicle was in motion. The time delay of T = 5 used in constructing phase space vectors y(n) is taken from the first minimum of the average mutual information. Results are shown for N = 87,000 data points.

SUMMARY, CONCLUSIONS, AND FUTURE DIRECTIONS

This paper has analyzed in some detail the pressure fluctuations measured on an axisymmetric body propelled under its own buoyancy at velocities on the order of 20 m/sec. At these velocities and with the thickness of the boundary layer about 0.5 cm, one establishes that the boundary layer thickness Reynolds number is about 100,000. In such flows there is substantial activity in the form of coherent structures [1, 2] whose size in 'wall length' units is about 100 wall units spanwise in the lower boundary layer and 150 to 200 units streamwise in the upper boundary layer.

These coherent structures are known to contribute in a significant way to the wall pressure fluctuations. The sensors used in this experiment have an effective sensing diameter of about 1400 in wall units. This is a very large sensor relative to the 'optimum' sensor size identified by Schewe [16] which is 19 wall units. The optimal size is established on several bases among which is proper evaluation of the whole set of degrees of freedom in the turbulent flow as seen in spectral analyses. The sensors in the present experiment certainly average over numerous short wavelength excitations of the fluid and sense scales which are consistent with those of the coherent structures. Indeed, on the basis of their size alone we would expect that these sensors would be responding to about ten or so coherent structures and report pressure fluctuations associated with the dynamics and interaction of this collection of coherent objects.

This qualitative picture is both consistent with and provides a rationale for the interpretation of the observations we have made in this paper when nonlinear signal processing methods [5] are applied to the data. These include:

- In the ambient data and in data taken from an accelerometer placed in the buoyant body we see that the mutual information 'decorrelation time' of the pressure fluctuations is a factor of two to three less than the times associated with fluid dynamical motions. The dimensions of the embedding space required to unfold the attractor in the dynamics is very large and is consistent with one's usual picture of 'noise'.

- In the data from the laminar flow regime of the boundary layer we also see time scales smaller than those of the fluid flow and see an embedding dimension which is very large. The laminar regime should have no pressure fluctuations, of course, so it is important to see this result as consistent with the general picture of the flow. The actual behavior of the global false nearest neighbors for these data suggests that it may be that there is a low dimensional dynamics (the beginning of Tollmien-Schlichting waves, perhaps) which is low amplitude in this region and is dominated by the ambient or instrumental 'noise' seen in the ambient and accelerometer data. It is impossible with just the geometric tool of false nearest neighbors to make a decision about this possibility.

- In the data in the transition region we see mutual information time scales which are consistent with the fluid dynamical excitations, namely about 240 µsec, and see low dimensional embedding dimensions chosen by the data itself as needed for the unfolding of the system attractor. The Fourier spectrum in this region is consistent with a band of Tollmien-Schlichting waves perhaps interacting with each other but now at substantially larger amplitude than might have been the case in the laminar region. The embedding dimensions required for the dynamics observed through the pressure sensors are about eight or so, varying with the individual sensor.

- In the data from turbulent flow in the boundary layer we see much the same pattern as in the transition region. The Fourier spectra have no distinguishable Tollmien-Schlichting region, as expected, and the time scale for mutual information decorrelation is slightly larger than in the transition region, though still clearly of fluid dynamic origin. The dimension required to unfold the attractor is about nine or ten in this flow regime.

We did perform some computations on the Lyapunov exponents for these data using methods well established in the literature [5] but are not yet confident enough of the numerical values to report them. The issue is the ability to quantitatively do computations on Lyapunov exponents in spaces of dimension eight to ten using the amount of data available. The qualitative picture which consistently emerges is that there is at least one positive Lyapunov exponent, as required by all descriptions of chaotic behavior, and one zero exponent demonstrating that the dynamics is that of a flow and not an iterated map. The sum of all exponents was always negative, so the flow is dissipative. Finally, the order of magnitude of the largest exponent was about $(10\ \tau_s)^{-1}$, with $\tau_s = 15.26$ µsec, the sampling time for these data. More details on these exponents will be reported in future work.

These analyses are remarkable in providing clear evidence for low dimensional dynamics within a flow which all agree is high dimensional when all degrees of freedom are accounted for. The numerical experiments of Keefe, Moin, and Kim [6] suggest that in flows of this sort with lower Reynolds numbers that dimensions (Lyapunov dimension) of 380 are seen. This rests on the full set of degrees of freedom excited by the flow. The present experiment has the good fortune to have utilized a sensor which averages out many of these degrees of freedom at the smallest scales and is sensitive to the dynamics and interaction of larger scale coherent structures in the boundary layer flow.

This raises the very interesting possibility for future experiments in this area: measure the embedding (or other) dimension as a function of sensor size (in wall units). If the explanation suggested here is qualitatively correct, we would expect that as the size of the sensor is increased the embedding dimension will slowly increase as more and more coherent structures are felt by the sensor. As the sensor size is decreased, the embedding dimension will decrease until the small scale motions become of importance and then the dimension will rise again. If this holds true, then for various applications of the observations here we would suggest using the minimum of such a dimension versus sensor size curve for operating purposes.

Further, this result suggests that the idea of what dimension one will see in an extended or continuum system with intrinsically many degrees of freedom depends on the resolution at which one looks. The strict notion suggested by the embedding theorem [3, 4, 12] that all degrees of freedom can be sensed by a single sensor of whatever size cannot be true physically unless the requirements of an infinite amount of infinitely accurate data are provided. There is a real challenge to the physical interpretation of experiments such as these to establish how dimension will vary as sensor resolution is varied. The same question arises in the area of analysis of climate and weather. Realistic resolution in measurements appropriate for that area is typically quite coarse. This means that the many, many degrees of freedom within the primitive equations for those flows are unlikely to be relevant to observations. The challenge of how to establish what is the relevant number of degrees of freedom as a function of resolution is what we pose here. The numerically based answer given in this paper is clearly only a bellwether that an interesting question is being posed.

The results in this paper suggest several directions of further inquiry:

- One should repeat the experiments both on the axisymmetric buoyant body and on appropriate laboratory flows. Instrumenting the body with pressure sensors of varying sizes is certainly called for as well as choosing the distribution of sensors to capture the spatial behavior of the turbulent boundary layer flows. The opportunity of using some flow visualization when possible could be a very useful contribution to our understanding of these flows and measurement of the velocity and vorticity within the boundary layer in the vicinity of the pressure sensors could be quite interesting.

- Using numerical simulations both existing (6) and higher resolution when possible would allow investigation of the dimension versus sensor size (effective spatial resolution) in these spatial-temporal chaotic settings. While no substitute for real experiment, numerical experiments here have clear and well tested advantages.

- Clarify the evaluation of local and global Lyapunov exponents (5) in the spaces of dimension eight to ten which one must deal with in these data. The predictability associated with both local and global exponents is important to establish. Further connecting large local exponents with properties of the coherent structures would be of high physical interest.

- Develop filtering methods (5) based on models, either locally deterministic or probabilistic, for removing the effect of the observed pressure fluctuations as they serve to mask external signals of interest.

- We anticipate the use of developments of existing 'control' methods for chaotic evolution [18] to allow control of the low dimensional dynamics seen in this experiment either to reduce the chaotic behavior, if that is desired, or to enhance it under other circumstances. The reduction would plausibly reduce 'drag', while the enhancement might allow the use of the chaos for steering or braking of objects moving rapidly in a fluid.

ACKNOWLEDGMENTS

HDIA thanks the members of INLS for numerous discussions on this subject. His work was supported in part by the U.S. Department of Energy, Office of Basic Energy Sciences, Division of Engineering and Geosciences, under contract DE-FG03-90ER14138, and in part by the Army Research Office (Contract DAAL03-91-C-052), and by the Office of Naval Research (Contract NOOO14-91-C-0125), under subcontract to the Lockheed/Sanders Corporation. Research at the Naval Undersea Warfare Center (NUWC) was performed under the technical direction of Dr. Richard A. Katz and was sponsored in part by the Science and Technology Directorate of NUWC, Dr. William Roderick, Director, (Special Project Initiative No. 793P13), and in part in collaboration with the USAF Foreign Aerospace Science and Technology Center, FASTC, Ms. Carolyn Scheaff, Director (Contract N4175693-WX-33012).

References

[1] Cantwell, B. J., "Organized Motion in Turbulent Flow", *Ann. Rev. Fluid Mech* **13** 457-515 (1981).

[2] Robinson, S. K., "Coherent Motions in the Turbulent Boundary Layer", *Ann. Rev. Fluid Mech.***23** 601-639 (1991).

[3] Mañé, R., in *Dynamical Systems and Turbulence, Warwick 1980*, eds. D. Rand and L. S. Young, *Lecture Notes in Mathematics* **898**, (Springer, Berlin), 230 (1981).

[4] Takens, F., in *Dynamical Systems and Turbulence, Warwick 1980*, eds. D. Rand and L. S. Young, *Lecture Notes in Mathematics* **898**, (Springer, Berlin), 366 (1981).

[5] Abarbanel, H. D. I., R. Brown, J. J. ("Sid") Sidorowich, and L. Sh. Tsimring, "The Analysis of Observed Chaotic Data in Physical Systems", to appear *Rev. Mod. Phys.*, October, 1993.

[6] Keefe, L., P. Moin, and J. Kim, "The Dimension of Attractors Underlying Periodic Turbulent Poiseuille Flow", *J. Fluid Mech.* **242**, 1-29 (1992).

[7] Galib, T. A. and A. Zadina, "Turbulent Pressure Fluctuation Measurements with Conventional Piezoelectric and Miniature Piezoresistive Transducers", NUSC Technical Memorandum 84-2045, 30 April 1984. (Unpublished)

[8] Gentry A. E. and A. R. Wazzan, "The Transition Analysis Program System, Volume II", McDonnell-Douglas Report No. MDC J7255/02, June, 1976. (Unpublished.)

[9] Galib, T. A., R. Katz, S. Ko, and B. Sandman, "Attenuation of Turbulent Pressure Fluctuations Through an Elastomeric Coating", NUSC Technical Memorandum 91-2083, 17 August 1991. (Unpublished).

[10] Farabee, T. M. and M. J. Casarella, "Spectral Features of Wall Pressure Fluctuations Beneath Turbulent Boundary Layers", *Phys. Fluids A* **3**, 2410-2420 (1991).

[11] Balint, J.-L., J. M. Wallace, and P. Vukoslavčević, "The Velocity and Vorticity Vector Fields of a Turbulent Boundary Layer. Part 2. Statistical Properties", *J. Fluid Mech.* **228**, 53-86 (1991).

[12] Eckmann, J.-P. and D. Ruelle "Ergodic Theory of Chaos and Strange Attractors" *Rev. Mod. Phys.* **57**, 617 (1985).

[13] Fraser, A. M. and Swinney, H. L., "Independent Coordinates for Strange Attractors" *Phys. Rev.*, **33A**, 1134-1140 (1986); Fraser, A.M., "Information and Entropy in Strange Attractors", *IEEE Trans. on Info. Theory*, 35, 245-262 (1989); Fraser, A. M., "Reconstructing Attractors from Scalar Time Series: A Comparison of Singular System and Redundancy Criteria" *Physica*, **34D**, 391-404 (1989).

[14] Kennel, Matthew B., R. Brown, and H. D. I. Abarbanel, "Determining Minimum Embedding Dimension using a Geometrical Construction", *Phys. Rev. A* **45**, 3403-3411 (1992).

[15] Abarbanel, H. D. I. and M. B. Kennel, "Local False Nearest Neighbors and Dynamical Dimensions from Observed Chaotic Data", Phys. Rev. E, 47, 3057-3068 (1993).

[16] Schewe, G., "On the Structure and Resolution of Wall-Pressure Fluctuations Associated with Turbulent Boundary-Layer Flow", *J. Fluid Mech.* **134** 311-328 (1983)

[17] Lorenz, E. N., "Deterministic Nonperiodic Flow" *J. Atmos. Sci.* **20**, 130 (1963).

[18] Ott, E., C. Grebogi, and J. A. Yorke, *Phys. Rev. Lett.* **64**, 1196 (1990).

INDEPENDENT VELOCITY INCREMENTS AND KOLMOGOROV'S REFINED SIMILARITY HYPOTHESES

G. Stolovitzky and K.R. Sreenivasan
Mason laboratory, Yale University
New Haven, CT 06520-2159

ABSTRACT

Under the assumption of statistical independence of velocity increments across scales of the order of the Kolmogorov scale, it is shown that a modified version of Kolmogorov's refined similarity hypotheses follows purely from probabilistic arguments. The connection of this result to three-dimensional fluid turbulence is discussed briefly.

1. INTRODUCTION

In 1962, Kolmogorov[1] put forward a refinement of his earlier phenomenological theory[2] of high-Reynolds-number turbulence. This refinement has become a vital reference point in the research of locally isotropic and homogeneous turbulence. An important quantity in this description is the flux of energy ϕ_r transferred across scales of size r. Kolmogorov assumed that, in the inertial range, ϕ_r is the only relevant quantity upon which all other quantities would depend. Furthermore, he identified ϕ_r with $r\varepsilon_r$, where ε_r is the rate of energy dissipation per unit mass averaged over a volume of linear scale r. Kolmogorov's theory is made quantitative on the basis of the following two celebrated hypotheses.

The first similarity hypothesis: If r«L, where L is a measure of the large scale of turbulence, the probability density function (pdf) of the stochastic variable

$$V = \frac{\Delta u(r)}{(r\varepsilon_r)^{1/3}} \qquad (1)$$

depends only on the local Reynolds number $Re_r = r(r\varepsilon_r)^{1/3}/\nu$, where ν is the kinematic viscosity of the fluid and $\Delta u(r) = u(x+r) - u(x)$, u being the x-component of the velocity vector $\mathbf{u}(x)$ and r is measured along x.

The second similarity hypothesis: If $Re_r \gg 1$, the pdf of V does not depend on Re_r either (nor on r, and is therefore universal).

Although it was shown recently[3] that the pdf of V depends on r as well, several aspects of these hypotheses have been verified experimentally[3,4] as well as by direct numerical simulations of turbulence[5]. Even before this verification, consequences of these hypotheses had been used extensively in the turbulence literature.

In spite of their widespread use, the hypotheses pose some troubling problems. For example, Kraichnan[6] has pointed out that, for r in the inertial range, $\Delta u(r)$ is a purely inertial range quantity, whereas $r\varepsilon_r$ is a mixed quantity (because ε_r is a dissipation quantity averaged over an inertial range scale), and so their ratio cannot be universal. Also, the notion of a cascade, where the energy is transferred locally in wavenumber space, has been criticized from time to time.

The primary shortcoming of Kolmogorov's hypotheses is that they have not yet been derived from basic principles. In this paper we wish to show that they can indeed be cast, under certain circumstances, in terms of general principles of stochastic processes. The physical picture of a cascade need not be assumed *a priori*, but rather as an *a posteriori* interpretation.

For convenience, we restrict our discussion to one-dimensional spatial cuts of the turbulent velocity fluctuation. In particular, we use the local isotropy approximation for the three-dimensional average dissipation rate, namely,

$$\varepsilon_r = 15\nu \frac{1}{r} \int_x^{x+r} \left(\frac{du}{dx}\right)^2 dx. \qquad (2)$$

Relaxing this assumption of local isotropy, Eq. (2), adds greater complexity to the proof given below, but it is believed that it will not affect its basic validity.

2. A CONVENIENT RESTATEMENT OF THE REFINED HYPOTHESES

The refined hypotheses are statements about the relation between the velocity increments

$$\Delta u(r) = \int_x^{x+r} \frac{du}{dx} dx \qquad (3)$$

and the energy dissipation rate in a segment of linear size r

$$r\varepsilon_r = 15\nu \int_x^{x+r} \left(\frac{du}{dx}\right)^2 dx. \qquad (4)$$

Given that both $\Delta u(r)$ and $r\varepsilon_r$ are functionals of the velocity gradient, they are (in general) correlated variables. Discretizing the integrals (3) and (4), and normalizing velocities by $(\eta\varepsilon)^{1/3}$ (where $\varepsilon = 15\nu \langle (du/dx)^2 \rangle$ and η is the Kolmogorov scale given by $(\nu^3/\varepsilon)^{1/4}$), and lengths by η, we may write (3) and (4) respectively as

$$S_p = \sum_{i=1}^{p} X_i \quad \text{and} \quad Y_p^2 = \sum_{i=1}^{p} X_i^2, \qquad (5)$$

where $X_i = \frac{du}{dx} \eta/(\eta\varepsilon)^{1/3}$ and $p=r/\eta$. Physically, the X_i represent normalized velocity increments across a distance η. In these variables, one can write that

$$V = S_p / Y_p^{2/3}, \tag{6}$$

and state Kolmogorov's similarity hypotheses as follows: the pdf of V depends only on $Re_r = pY_p^{2/3}$ and the variable p, and becomes independent of both when $p \gg 1$ and $Y_p \gg 1$.

The discretized equations (5) suggest that a more general formulation of the refined similarity hypotheses in terms of stochastic processes might be attainable. This is done in the next section.

3. REFINED SIMILARITY HYPOTHESES FOR BROWNIAN MOTION

Let us assume that the X_i are normally distributed independent random variables with zero mean and variance σ^2. We then have the following exact result.

Theorem: Given that $p \geq 2$ is an integer, the pdf of $\xi = S_p/Y_p^{2H}$ conditioned on Y_p is independent of Y_p only when $H=1/2$. In such a case, the conditional pdf assumes the form:

$$f(\xi|p, Y_p) = \frac{1}{\sqrt{p}\Omega_\theta(p)} \left(1 - \frac{\xi^2}{p}\right)^{(p-3)/2}, \quad \xi^2 < p \tag{7}$$

where $\Omega_\theta(p) = \frac{\pi}{2^p} \frac{\Gamma(p-1)}{\Gamma(p/2)^2}$, $\Gamma(x)$ being the gamma function.

Proof: First we compute the joint pdf of S_p and Y_p for a given p as

$$P_2(S_p, Y_p|p) = $$
$$\int_{R^p} dX_1...dX_p \Pi_X(X_1,...,X_p) \delta(S_p - \sum_{i=1}^{p} X_i) \delta(Y_p - (\sum_{i=1}^{p} X_i^2)^{1/2}) \tag{8}$$

where R^p is p-dimensional real space, $\delta(x)$ is Dirac's delta function, $\Pi_X(X_1,..., X_p)$ is the joint probability of X_i and is equal to $(2\pi\sigma^2)^{-p/2} \exp(\sum_{i=1}^{p} X_i^2/2\sigma^2)$. To evaluate the integral we perform an orthogonal change of coordinates from $X \to U$:

$$U_j = \sum_{i=1}^{p} R_{ij} X_j, \tag{9}$$

where R_{ij} is an orthogonal matrix with $R_{1j} = 1/\sqrt{p}$. The other rows of this matrix, R_{kj}, $2 \leq k \leq p$, can be computed using (for example) a Gram-Schmidt procedure, but they are irrelevant for our purposes here. As the Jacobian of this transformation is unity, the integral (8) can be written as

$$P_2(S_p, Y_p | p) = \int_{R^p} dU_1 \ldots dU_p \, \Pi_U(U_1, \ldots, U_p) \, \delta(S_p - \sqrt{p}\, U_1) \, \delta(Y_p - (\sum_{i=1}^{p} U_i^2)^{1/2}), \quad (10)$$

where $\Pi_U(U_1, \ldots, U_p) = \Pi_X(X_1(U_1, \ldots, U_p), \ldots, X_p(U_1, \ldots, U_p))$.

The next step is to change to p-dimensional spherical polar coordinates:

$$U_1(\rho, \theta, \phi_1, \ldots, \phi_{p-2}) = \rho \cos\theta$$
$$U_2(\rho, \theta, \phi_1, \ldots, \phi_{p-2}) = \rho \sin\theta \cos\phi_1$$
$$U_3(\rho, \theta, \phi_1, \ldots, \phi_{p-2}) = \rho \sin\theta \sin\phi_1$$
$$U_4(\rho, \theta, \phi_1, \ldots, \phi_{p-2}) = \rho \sin\theta \sin\phi_1 \cos\phi_2$$
$$\ldots$$
$$U_{p-1}(\rho, \theta, \phi_1, \ldots, \phi_{p-2}) = \rho \sin\theta \sin\phi_1 \ldots \sin\phi_{p-3} \cos\phi_{p-2}$$
$$U_p(\rho, \theta, \phi_1, \ldots, \phi_{p-2}) = \rho \sin\theta \sin\phi_1 \ldots \sin\phi_{p-3} \sin\phi_{p-2} \quad (11)$$

where $0 < \rho < \infty$, $0 < \theta < \pi$, $0 < \phi_i < \pi$ for $i=1, \ldots, p-3$ and $-\pi < \phi_{p-2} < \pi$. Noting that the Jacobian of this transformation is

$$\frac{\partial(U_1, \ldots, U_p)}{\partial(\rho, \theta, \phi_1, \ldots, \phi_{p-2})} = \rho^{p-1} (\sin\theta)^{p-2} (\sin\phi_1)^{p-3} \ldots (\sin\phi_{p-3}), \quad (12)$$

we can perform the integral in Eq. (10) in polar coordinates, to yield

$$P_2(S_p, Y_p | p) = \Omega_\phi(p) \left(1 - \frac{S_p^2}{p Y_p^2}\right)^{(p-3)/2} \frac{\exp(-Y_p^2/2\sigma^2)}{(2\pi\sigma^2)^{p/2}}, \quad (13)$$

where $\Omega_\phi(p) = \frac{\pi(p-1)}{\Gamma((p-1)/2)}$ is the integral over ϕ_1 through ϕ_{p-2} of $(\sin\phi_1)^{p-3} \times \ldots (\sin\phi_{p-3})$ arising from the Jacobian (12). The probability of S_p conditioned on Y_p is computed as

$$P(S_p | p, Y_p) = \frac{P_2(S_p, Y_p | p)}{\int dS_p P_2(S_p, Y_p | p)}, \quad (14)$$

and we find

$$P(S_p|p,Y_p) = \frac{1}{\sqrt{p}Y_p\Omega_\theta(p)}\left(1 - \frac{S_p^2}{pY_p^2}\right)^{(p-3)/2}. \quad (15)$$

where $\Omega_\theta(p) = \int_0^\pi (\sin\theta)^{p-2}$ and is expressed in a closed form in the statement of the theorem. If we now change variables to $\xi = \frac{S_p}{Y_p^{2H}}$, we obtain

$$f_H(\xi|p,Y_p) = \frac{1}{\sqrt{p}Y_p^{1-2H}\Omega_\theta(p)}\left(1 - \frac{\xi^2}{pY_p^{2-4H}}\right)^{(p-3)/2}. \quad (16)$$

It is easily seen that the only value of H that renders $f_H(\xi|p,Y_p)$ independent of Y_p is H=1/2, and the form of $f_{1/2}(\xi|p,Y_p)$ is then the one stated in the theorem.

In fact, the theorem can be stated to appear closer to Kolmogorov's hypotheses:
H1) The pdf of the stochastic variable $\xi = S_p/Y_p$ conditioned on Y_p depends only on p.
H2) When p»1, the pdf of ξ becomes independent of p; in fact, it tends to the normal distribution.

Plots of $f(\xi|p,Y_p)$ for different values of p are shown in Fig. 1. These functions are supported in the interval $(-\sqrt{p},\sqrt{p})$. It can be seen that the distribution for p=2 is bimodal. The distribution for p=3 is uniform between $\sqrt{3}$ and $-\sqrt{3}$. For larger p,

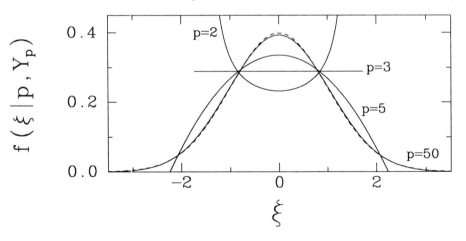

Fig. 1. Plots of $f(\xi|p,Y_p)$ for different values of the parameter p when the velocity increments X_i are assumed to be Gaussian.

the trend towards a Gaussian distribution occurs; for p=50 the departure from the Gaussian shape (dashed line) is negligible.

4. MORE GENERAL PROCESSES

It is useful to explore the consequences of relaxing the hypotheses that the X_i are normally distributed, and that they are independent. We will address the former issue here, and leave the latter for a forthcoming publication[7].

Let $\xi = S_p/Y_p$ and the X_i be independent random variables with zero odd-order moments and non-divergent even-order moments. From these conditions it is clear that for odd n, $\langle \xi^n | p, Y_p \rangle = 0$. We wish to show that for $p \gg 1$ and even $n \geq 2$

$$\langle \xi^n | p, Y_p \rangle \approx (n-1)!!, \quad (17)$$

which implies that ξ is a normal random variable with zero mean and unity variance. In particular, for $p \gg 1$ the pdf of ξ is independent of p and Y_p. We give an informal proof of this result below.

The n-th power of S_p can be written as

$$S_p^n = \sum_{i_1+\ldots+i_p=n} \frac{n!}{i_1!\ldots i_p!} X_1^{i_1} \ldots X_p^{i_p} \quad (18)$$

where the sum extends over all the possible $0 \leq i_j \leq n$ such that $\sum_1^p i_j = n$. It is clear from Eq. (18) that, if one computed $\langle S_p^n \rangle$, all terms containing odd i_j will not contribute to the sum. This follows from the independence of the X_i and the assumption that odd-order moments of X_i are zero. Further, when $p \gg 1$, most of the contribution to S_p^n is likely to come from terms containing even i_j because the terms with odd i_j (which, in general, possess both signs) are likely to cancel each other. Among all the terms with even i_j, the most numerous ones are those for which $\frac{n!}{i_1!\ldots i_p!}$ takes a maximum value. This happens when $i_j=0$ or 2, and the number of such terms will be $\binom{p}{n/2} \sim p^{n/2}$. The next most numerous terms are those for which $i_j=0$, 2 or 4 (with 4 occurring only once), and there will be $\binom{p}{n/2-1} \sim p^{n/2-1}$ such terms. If all the terms are of the same order, it is clear that the terms with $i_j=0$ or 2 dominate the sum, so that we may rewrite Eq. (18) as

$$S_p^n \approx \sum_{i_1+\ldots+i_p=n} \frac{n!}{2^{n/2}} X_1^{i_1} \ldots X_p^{i_p}, \quad (19)$$

where i_j takes only the values 0 or 2. To make this statement more rigorous, it would be necessary to bound the errors made in this approximation, but this will not be attempted here.

We can make a similar analysis for Y_p and obtain, with i_j assuming only values of 0 or 1 in Eq. (20a) and values of 0 or 2 in Eq. (20b), that:

$$Y_p^n \approx \sum_{i_1+...+i_p=n/2} (n/2)!\, X_1^{2i_1} ... X_p^{2i_p}, \quad (20a)$$

$$\approx \sum_{i_1+...+i_p=n} (n/2)!\, X_1^{i_1} ... X_p^{i_p}. \quad (20b)$$

By comparing Eqs. (19) and (20b), we obtain

$$\langle S_p^n | Y_p \rangle \approx \frac{n!}{2^{n/2}(n/2)!} Y_p^n. \quad (21)$$

Noting that $\dfrac{n!}{2^{n/2}(n/2)!} = (n-1)!!$, we can restate Eq. (21) as

$$\langle \xi^n | Y_p \rangle \approx (n-1)!!, \quad (22)$$

for $p \gg 1$, as required. We thus conclude that in this limit the pdf of ξ conditioned in Y_p is independent of p and Y_p. Actually, it tends to be Gaussian with zero mean and unity variance.

We now illustrate these results numerically. For definiteness, we will consider that the X_i are distributed according to an exponential density g, i.e.

$$g(X) = \frac{1}{2} \exp(-|X|). \quad (23)$$

For this distribution, $\langle X^2 \rangle = 2$, and hence $\langle Y_p \rangle = 2p$. We computed the pdfs $h(\xi|p,Y_p)$ of $\xi = \dfrac{S_p}{Y_p}$ conditioned on Y_p for different values of p. The values of the conditioning parameter Y_p are taken as windows of size $\langle Y_p \rangle/3$ centered at the values indicated in Figs. 2. We see from Fig. 2a that the distributions for p=3 are bimodal and show some dependence on Y_p. Recall that for p=3, the equivalent distribution for the case of Gaussian X_i is uniform; we conclude that the conditional distributions of ξ for small p do depend on the distribution of X_i. For p=10 (Fig. 2b), the distributions exhibit Gaussian-like behavior while for p=50 (Figs. 2c and 2d) they differ very little from the Gaussian. As p increases, the differences among the various curves corresponding to different values of Y_p tend to diminish. In particular, for p=50, all four curves coalesce quite well.

The behavior just discussed can be described in the same terms as the Kolmogorov's similarity hypotheses: the pdf of ξ conditioned on Y_p depends on Y_p and p. For $p \gg 1$, it tends to a Gaussian with zero mean and unity variance. This behavior is independent of the particular distribution one chooses for X, although details for small values of p do depend on this choice.

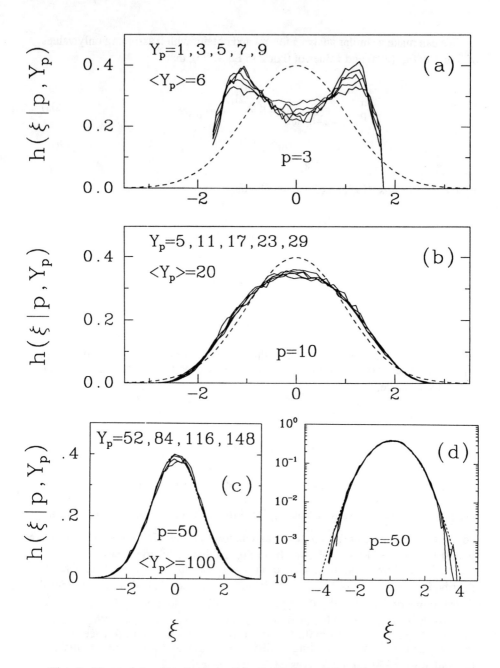

Fig. 2: Plots of the distribution of $\xi = S_p/Y_p$ conditioned on Y_p, $h(\xi|p,Y_p)$, for different values of the parameter p, when the X_i are exponentially distributed. (a) p = 3, (b) p = 10, (c) p = 50, and (d) p = 50, with logarithmic ordinate. The dashed lines in each of these figures is the normal density.

A final remark is in order. Even though this result looks similar to the central limit theorem, its says something more. While it is not difficult to derive the central limit theorem from it, as far as we are aware, this result cannot be derived from the central limit theorem.

5. CONCLUSIONS

We have seen that Kolmogorov's hypotheses can be phrased in a broader context than envisaged originally. The arguments leading to the hypotheses are taken from probability theory, and do not invoke any physical picture such as the cascade of energy. However, it is necessary to clarify the connection to the physics when dealing with real turbulence data. First of all, it is clear that in real turbulence data there exist correlations between the X_i. Therefore, the results discussed in this paper do not apply to turbulence in a straightforward fashion. In correlated processes where the X_i are not independent, the exponent H in

$$S_p = V Y_p^{2H} \qquad (24)$$

will be different from 1/2. For turbulence we have to go to the Navier-Stokes equations to find the value of H. In the present notation, Kolmogorov's equation for the third-order structure function in the inertial range, $<\Delta u(r)^3> = -(4/5)r\varepsilon$, which is derived from Navier-Stokes equations with the additional assumptions of local isotropy and homogeneity, states that

$$<S_p^3> = -\frac{4}{5}(15^{1/3}) <Y_p>. \qquad (25)$$

If we assume that V is independent of Y_p for $p \gg 1$, we are led to the choice H=1/3. Equation (24) is in this case the same as Eq. (6), and Kolmogorov's hypotheses follow. This also implies that $<V^3>$ is different from zero, and hence, the limiting pdf of V for $p \gg 1$ is not a Gaussian. The detailed study of these issues for correlated systems - in particular for the fractional Brownian motion - and its comparison with atmospheric turbulence will be published elsewhere[7].

Acknowledgements: This work was supported by AFOSR and DARPA.

REFERENCES

1. A. N. Kolmogorov, J. Fluid Mech. **13**, 82 (1962)
2. A. N. Kolmogorov, Dokl. Akad. Nauk SSSR **30**, 301 (1941)
3. G. Stolovitzky, P. Kailasnath and K. R. Sreenivasan, Phys. Rev. Lett. **69**, 1178 (1992)
4. A. A. Praskovsky, Phys. Fluids A **4**, 2589 (1992); S. T. Thoroddsen and C. W. Van Atta, Phys. Fluids A **4**, 2592 (1992);
5. S. Chen, G. D. Doolen, R. H. Kraichnan and Z-S. She, Phys. Fluids A **5**, 458 (1993); I. Hosokawa, J. Phys. Soc. of Japan, **62**, 10 (1993)
6. R. H. Kraichnan, J. Fluid Mech. **62**, 305 (1974)
7. G. Stolovitzky and K. R. Sreenivasan, preprint.

PROCESSING OF MEASURED TRANSITIONAL AND TURBULENT TIME SERIES

John Salisbury
Analysis & Technology
Engineering Technology Center
Newport, R.I.

Thomas A. Galib
Naval Undersea Warfare Center
Newport, R.I.

ABSTRACT

This analysis investigated experimentally the chaotic and spectral dynamics of an open boundary layer in the direction of flow to acquire a better understanding of the noise generating mechanisms. If the mechanisms are low dimensional chaotic processes, a variety of chaotic noise reduction techniques would be available. The specific data parameters investigated were: time series of wall pressure disturbance, temporal spectra, autocorrelation, embedding dimension, lyapunov exponent, correlation dimension, and phase space trajectories. Evidence of low dimension dynamics was found in the impingement wall pressure time series data. The major system test parameters were: vehicle velocity,41 knots; sampling rate, 65536 Hz; and filter bandwidth, low pass @ 6400 Hz. The data conditioning was performed in the Naval Undersea Warfare Center, Test and Evaluation Signal Processing Facility.

EXPERIMENT

The experiment was conducted in a deep fresh water lake, Lake Pend Oreille, at a facility operated by the David Taylor Research Center/Acoustic Research Detachment in Bayview, Idaho. A 21 inch diameter buoyant test vehicle was used for the experiment. The test vehicle was propelled vertically from a depth of 1100 feet near the bottom of the lake by its buoyancy (Figure 1).

The buoyant test vehicle was a 21 inch diameter axisymmetric body approximately 27 feet in length. The vehicle included a weight section which held a symmetrically distributed array of cylindrical lead weights. Depending upon the amount of weight in this section, vehicle speeds of 40 to 75 feet per second were attainable. A speed of 70 feet per second was used for the experiment described herein.

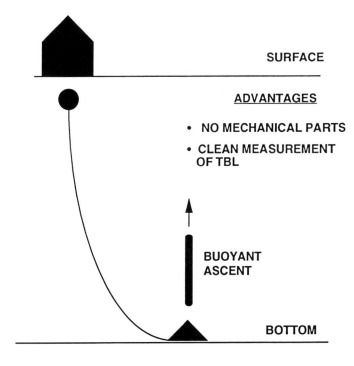

Figure 1. Buoyant Test Vehicle Experimental Setup

The nose of the test vehicle (Figure 2) was instrumented with piezoelectric pressure transducers, in order to measure the pressure fluctuations associated with the developing boundary layer and the resulting turbulence. Note: The artist's depiction of transition in Figure 2 is not meant to imply that 50% intermittency occurs halfway through the transition region. Also, the boundary layer is growing between δ_L and δ_T as spots travel through this region. The transducers were PCB model 112M149 having a sensitivity of 50 mV/psi (-26dB/1V/psi). The physical diameter of the transducers was 0.218 inch, and the effective diameter, based on -6 dB sensitivity, was determined to be 0.12 inch[1]. The pressure transducers were mounted in a carbon-graphite nose shell, with stainless steel inserts, as shown in Figure 3. The entire shell surface was covered with a 0.125 inch elastomer in order to provide a smooth surface for the boundary layer. Thus, in this configuration, the transducers did not interfere with the flow over the nose surface.

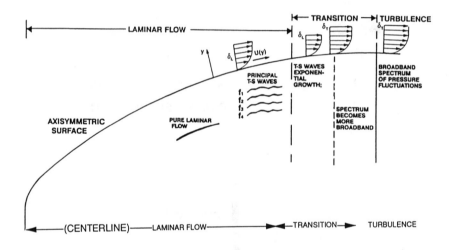

Figure 2. Regions of Laminar, Transitional, and Turbulent Flow Along an Axisymmetric Cylindrical Body

The vehicle was hauled down to a depth of 1100 feet via a cable attached to an onshore winch. Once the vehicle was stopped at the bottom of the lake, the onboard Honeywell 5600C tape recorder was powered, and ambient data from all sensors was recorded for 30 seconds. Following the ambient recording, the vehicle released itself from the cable and buoyantly ascended toward the surface. For a vehicle speed of 70 feet per second, steady state conditions were achieved at a depth of 700 feet. At a depth of 160 feet, the fins "kicked over," turning the vehicle so that it would not broach the surface. The data were recorded during the steady state portion of the run, in this case lasting for about five (5) seconds while traversing between 550 and 200 feet.

The time averaged spectra of pressure fluctuations measured by transducers in laminar, transitional, and turbulent flow are show in Figures 4-8. Length Reynolds number and displacement thickness are also shown in the figures. The displacement thickness was computed by the Transition Analysis Program System[2]. Because there are no measurable pressure fluctuations in laminar flow, the spectrum for laminar flow consists of the transducer response to fluid-loaded nose vibration, and not to boundary layer pressure fluctuations. This is an important point to be remembered for the results of this study. Transition and turbulence data are dominated by boundary layer pressure fluctuations up to 3 kHz[3].

CHARACTERISTICS OF CHAOS

Selected diagnostics used in this study are from the field of Chaos[4], and have the following characteristics:

- Sensitivity to Initial Conditions
- Finite Dimension (Fractal)
- Positive Lyapunov Exponent
- Broad Spectrum
- Phase Space Trajectory (Strange Attractor)

A dynamical system having these features, is a chaotic system and if the embedding dimension is low, <10, then the system is deterministic and can be modeled using a few differential equations. A random signal will have an embedding dimension of infinity. The ability to model a dynamic system gives us some predictive capability and therefore techniques for noise reduction.

When a dynamic system is said to have Sensitivity to Initial Conditions (SIC), it implies that the separation of the trajectories of two points, initially very close, will diverge exponentially as the dynamic system is evolved.

110 Measured Transitional and Turbulent Time Series

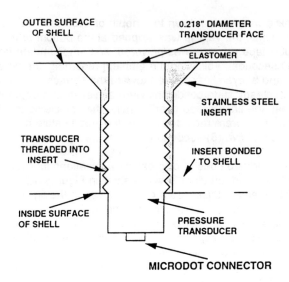

Figure 3. Typical Transducer Configuration

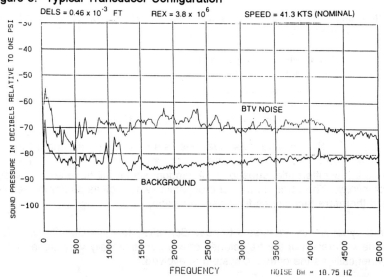

Figure 4. Temporal Power Spectrum of: (a) Wall Pressure Fluctuations In Laminar Region of Flow (Station B1) and, (b) Ambient Background

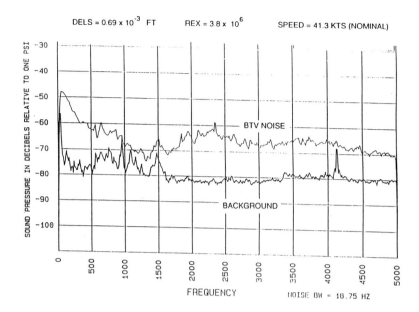

Figure 5. Temporal Power Spectrum of: (a) Wall Pressure Fluctuations In Laminar Zone (Station B3) and, (b) Ambient Background

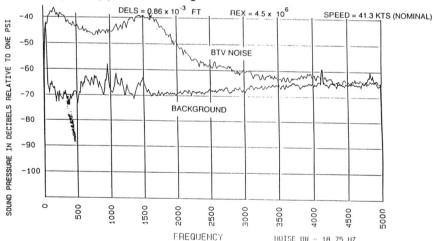

Figure 6. Temporal Power Spectrum of: (a) Wall Pressure Fluctuations in Transition Zone (Station B4) and, (b) Ambient Background

112 Measured Transitional and Turbulent Time Series

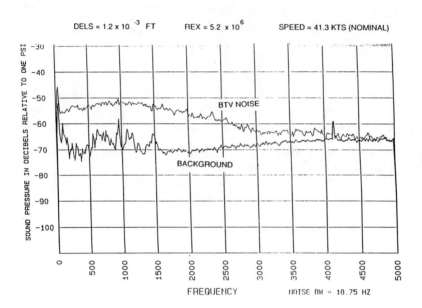

Figure 7. Temporal Power Spectrum of: (a) Wall Pressure Flucuations in Turbulence Zone (Station B5) and, (b) Ambient Background

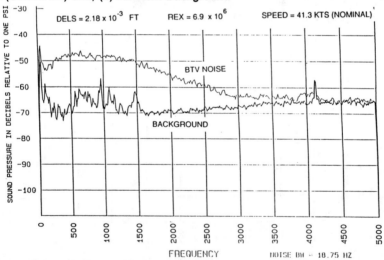

Figure 8. Temporal Power Spectrum of: (a) Wall Pressure Fluctuations in Turbulence Zone (Station B7) and, (b) Ambient Background

The SIC concept is related to the Lyapunov exponents. These exponents measure the exponential attraction or separation in time of two adjacent trajectories in phase space with different initial conditions. A positive Lyapunov exponent indicates a Chaotic motion in a dynamic system with bounded trajectory. Although the temporal power spectrum obtained from a chaotic time series exhibits a broadband continuum akin to random-like signals, the phase space trajectory yields a deterministic topology.

The points in the bounded trajectory, are sometimes referred to as a 'strange attractor',whose dimension is finite and fractal. This is a quantitative property of a set of points in a 'd' dimensional space which measures the extent to which the points fill a subspace as the number of points becomes very large. There are several techniques for calculating the fractal dimension: Hausdorff, Correlation Integral and Local Intrinsic Dimension. A computationally efficient method for determining the minimum embedding dimension of a signal of interest is the False Neighbor Technique, which we incorporate in the following section.

DATA ANALYSIS

Six data sets were selected from the test and all have been analyzed. The results of only three representative sets will be discussed here, B1 (Laminar Region), B4 (Transitional Region), and B7 (Turbulence Region). The time history of the data sets are presented in Figures 9, 10, and 11. The scale is the same in all figures in order to show relative levels in all regions: low level for B1, high level for B4, and intermediate for B7. The signal-to-noise ratio (SNR) in all cases was greater than 10 dB.

In order to perform some of the Chaotic analysis, the scaler time data must be embedded into a 'd' dimension vector field. One technique for accomplishing this is by the method of delays. That is to say, a vector is generated by selecting 'd' data points, each separated by a delay of t samples. The delta t can be selected in several ways. The most common is to select the zero crossing of the autocorrelation function or the first minimum of the mutual information function. Figures 12, 13, and14 show the autocorrelation functions of B1, B4, and B7 respectively. The plots disclose that the dynamic systems are not purely random, but there exist some structure in the data. The unexplained peak at sample 32 in Figure 12 appears as a 1920 Hz tonal in the data. The peak in Figure 13 is due to a tonal at 1480 Hz. This is the frequency region of the Tollmien Schlichting (TS) wave for the given speed and geometry of the boundary layer. Again, for Figure 14, there exist a 1213 Hz tonal. The delay time 't', in samples, selected from these figures are; 7, 16, and 18 respectively. The mutual information for a 2 dimension case was computed for this data and resulted in the same sample delays.

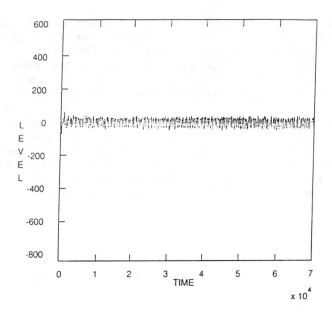

Figure 9. Time History at Station B1 (Laminar Zone)

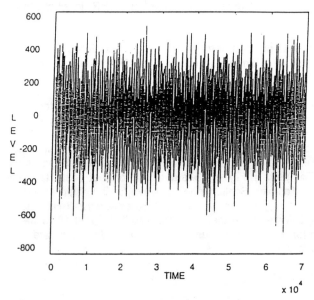

Figure 10. Time History At Station B4 (Transition Zone)

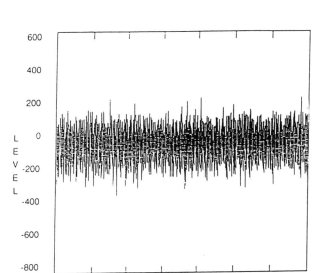

Figure 11. Time History At Station B7 (Turbulence Zone)

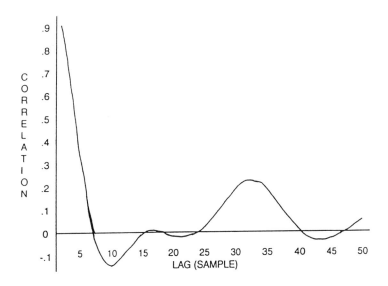

Figure 12. Autocorrelation For Station B1 Time Series Measurement

116 Measured Transitional and Turbulent Time Series

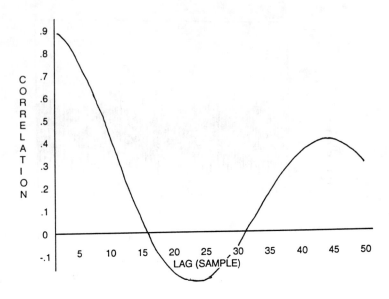

Figure 13. Autocorrelation For Station B4 Time Series Measurement

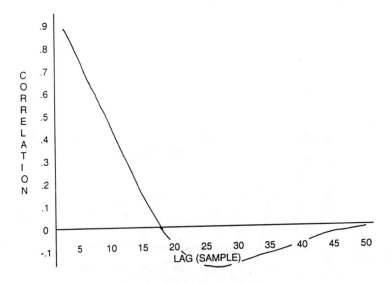

Figure 14. Autocorrelation For Station B7 Time Series Measurement

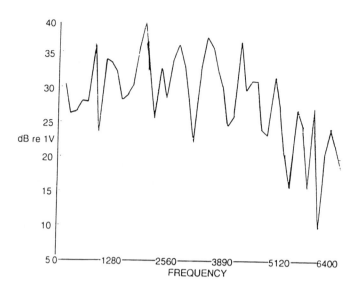

Figure 15. Short Term Power Spectrum, Station B1

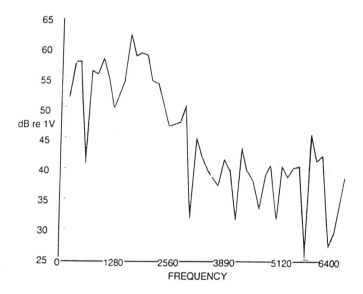

Figure 16. Short Term Power Spectrum, Station B4

An analysis of the embedding dimension was performed using the False Neighbor Technique. It was determined that for transition zone station measurements, a solution convergence to 0% false neighbor occurred at integer dimension value of 7, and for turbulence zone station measurements a convergence to 0% occurred at about dimension 8.

Figures 15, 16, and 17 show single snapshots of the frequency spectrum, with no averaging as was done previously. Figure 15 displays a broad band signal having a maximum value at approximately 1920 Hz. This value agrees with the autocorrelation results. Figure 16 depicts the transition region. It is in this region that the TS waves are generated and for this geometry, the TS waves are observed at a frequency of 1480 Hz, and the broad band level is above (approximately 20 dB) that in the laminar region (Figure 15). Also present are the harmonics of the TS wave, 2960, and 4440 Hz. Figure 17 is the spectrum of the Turbulent Region. Again, there is a broadband region dropping off rapidly at 1760 Hz and the tonals in the 1200 Hz region as predicted by the autocorrelation function.

Phase Plots were generated by plotting L_i vs L_{i+11} where L_i is the 'i th' sample level and L_{i+11} is the 11th sample level after L_i. Figures 18, 19, and 20 depict the three phase plots respectively using 2000 data points for the three corresponding data sets (B1, B4, B7). The shift of 11 samples was selected because it represents a time shift of a quarter of the TS wave period. These plots give an indication of the relative size of the dynamic systems, as do the relative levels in the time history plots. A phase plot of a random signal would be displayed as a circular glob of data points. The definite circular structure in Figure 19 represents the period of the TS wave. The phase plot in Figure 20 also depicts deterministic structure, of relatively low order as compared to the phase plot in Figure 18.

A Chaotic dissipative dynamic system has at least one positive Lyapunov exponent. This indicates that the system is exponentially expanding. The Lyapunov exponent was calculated for all the data sets using the WSSV technique[6]. This technique computes only the largest exponent and there are questions regarding its full capabilities. Figures 21, 22, and 23 present the results of the computation for B1, B4, and B7. All the data sets exhibit a positive Lyapunov exponent, indicating Chaotic systems.

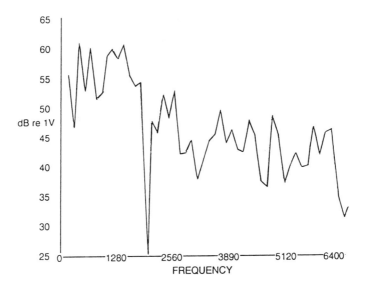

Figure 17. Short Term Power Spectrum, Station B7

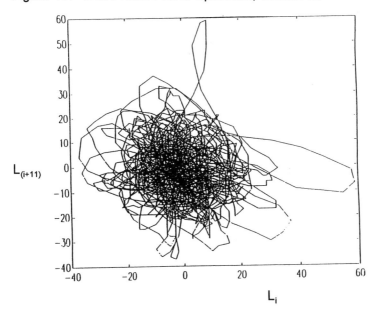

Figure 18. Phase Plot, Station B1

120 Measured Transitional and Turbulent Time Series

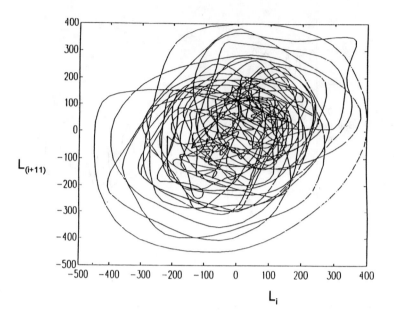

Figure 19. Phase Plot, Station B4

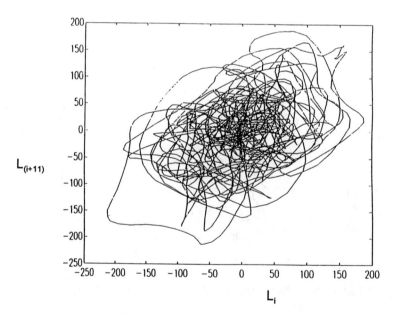

Figure 20. Phase Plot, Station B7

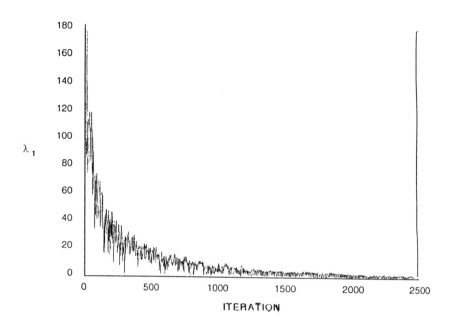

Figure 21. Lyapunov Exponent, Station B1

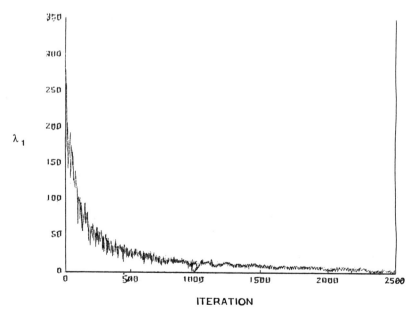

Figure 22. Lyapunov Exponent, Station B4

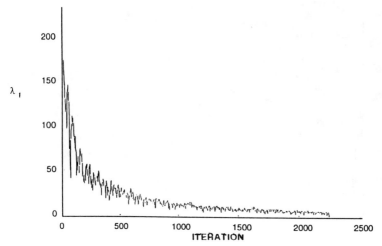

Figure 23. Lyapunov Exponent, Station B7

SUMMARY OF RESULTS AND CONCLUSIONS

The following table summarizes the results obtained from chaotic data processing:

Sensor	Zone	Time Delay	Embedding Dimension	Lyapunov Exponent
B1	Laminar	7	(Non-Convergent)	+2
B4	Transition	16	7	+4
B7	Turbulent	18	8	+5

The low finite embedding dimension (ED), which was calculated using the False Neighbor technique, is a very desirable result, since it implies that the boundary layer can be modeled. This capability gives us some control over its behavior and thereby noise reduction techniques can be applied.

Preliminary conclusions resulting from this study include:

- The dynamics of the boundary layer do have chaotic characteristics
- There appear to be the following two mechanisms involved in the generation of the flow characteristics:
  1) Transition region: nonlinear, chaotic
  2) Turbulence region: nonlinear, chaotic
- Pressure fluctuations in the laminar zone are attributed to low level vibrational noise.
- The TS waves from the transition region generate harmonics and are evident in the turbulent region
- More analysis is needed to generate models of the mechanisms

REFERENCES

1. Galib, T.A. and Sadina, A. "Turbulent Pressure Fluctuation Measurements with Conventional Piezoelectric and Miniature Piezoresistive Transducers," NUSC TM 84-2045, 30 April 1984.

2. Gentry, A. E. and Wassan, A. R., "The Transition Analysis Program System, Volume II," McDonnell-Douglas Report No. MDC J7255/02, June 1976.

3. Galib, T. A., Katz, R., Ko, S., and Sandman, B. "Attenuation of Turbulent Pressure Fluctuations Through an Elastomeric Coating," NUSC TM 912083, 17 August 1991.

4. Moon, F. C., "Chaotic Vibrations; An Introduction for Applied Scientists and Engineers," John Wiley & Sons, 1987.

5. Wolf, A., Swift, J. B., Swinney, H. L., and Vastano, J., "Determining Lyapunov Exponents from a Time Series," Physica, 16D, 285-317, 1985.

SYNCHRONIZATION AND CONTROL OF CHAOS

SYNCHRONIZING CHAOTIC CIRCUITS

Thomas L. Carroll and Louis M. Pecora
Naval Research Laboratory, Washington, DC 20375

ABSTRACT

Recent work using chaotic signals to drive nonlinear systems shows that chaotic dynamics is rich in new application possibilities. The approach of using nonlinear dynamics concepts to guide synthesis of new nonlinear systems leads to the concept of synchronization of chaotic systems. We demonstrate this concept here in computer models and in circuits. We also show how these ideas might be used for communications.

1. NEW DRIVING SIGNALS

Driven systems are easily visualized as dynamical systems which have as one of their parameters a dynamical variable from another, often autonomous, dynamical system. We often refer to the source of the driving signal as the *drive* system and to the driven system as the *response* system. This is a one-way setup in which the response "feels" the input of the drive, but the drive is unaware of the response. Typically, when driven systems are studied or engineered the driving signals come from either fixed point systems (e.g. constant forces) or limit cycles (e.g. sine wave forcing). The use of signals from a chaotic system to drive a nonlinear system offers a new type of driving signal – one not yet studied in any depth approaching sinusoidal forcing.

Like sinusoidal signals chaotic signals have their own unique properties. Their effects on driven systems will likewise be unique. If chaotic behavior is to become something useful, rather than merely something to be avoided, chaotically driven systems need to be studied. This is our motivation for studying chaotically driven systems.

In our approach two major themes stand out. One is the idea of stability as generalized to chaotic systems. This involves the use of Lyapunov exponents and generalizes in a straightforward way many properties of stable systems. Nevertheless these simple generalizations lead to some unexpected, but interesting and potentially useful behavior.

Another theme is the use of a constructive approach to building useful, chaotically driven systems. We cut apart, duplicate, and paste together nonlinear dynamical systems. Some of this necessitates trial and error since, as we mentioned, chaotic driving is rather new. But many things can be done with some guidance from what is now known in nonlinear dynamics. This approach is exciting in both an experimental and a theoretical way, the latter because it often opens up new mathematical questions which do not occur in sinusoidally driven systems. We will list some of these later.

We will study the first theme of stability as a prelude to the synchronization of chaotic circuits.

2. STABILITY OF CHAOTICALLY DRIVEN SYSTEMS

Consider a general nonlinear response system, $\dot{w} = h(w,v)$, where the driving signal v is supplied by a chaotic system. The question of stability arises when we ask: given a trajectory w(t) generated by this system for a particular drive v, when is w(t) immune to small differences in initial conditions, i.e. when is the final trajectory unique, in some sense? Fig. 1 shows this schematically.

To appreciate how this applies to chaotically driven responses, consider two responses w and w' started at slightly different points in phase space (w'(0)=w(0)+Δw(0)). We want to know the conditions under which w'(t) $\to$ w(t) as t$\to\infty$. Assume that Δw(0) is small and subtract the equations of motion for w' and w. This gives an equation of motion for Δw:

$\Delta\dot{w}=D_w h(w,v) + o(w,v)$, where $D_w h$ is the Jacobian of h with respect to w and o represents the remainder of the terms. The question of stability is now when does $\Delta w \to 0$? See Fig. 1 for a picture of this and Ref. [1] for more details.

Typically, a linear stability analysis invoked in this case employs the fundamental theorem of linear stability:

Theorem. The null solution (x=0) of the nonlinear non-stationary system $\dot{x}=A(t)x + o(x,t)$, with $o(0,t) = 0$ for all t, is uniformly asymptotically stable if: (i)

$$\lim_{\|x\|\to 0} \frac{\|o(x,t)\|}{\|x\|} = C \qquad (1)$$

holds uniformly with respect to t; (ii) A(t) is bounded for all t; (iii) the null solution of the linear system $\dot{x}=A(t) x$ is uniformly asymptotically stable.

What this means is that if (i)-(iii) are true, then there will always be a non-empty set of initial conditions which will asymptotically converge to zero. This compares directly with the equations for the response differences by replacing x by Δw, A by $D_w h$, and realizing that v just adds a time dependence to o and $D_w h$.

For most dynamical systems (i) and (ii) are easily satisfied. Now if we can establish (iii), we have the results that, despite a chaotic drive, the w system will always settle into the same trajectory and the points on that trajectory will eventually be at the same place at the same time for any initial condition in an open set around w(0).

For fixed point behavior part (iii) is established by examining the eigenvalues of the Jacobian $D_w h$ and for sinusoidal driving one uses Floquet multipliers, but our case is neither of these and we must use something more general and suitable for chaotic behavior. These are the Lyapunov exponents of the w system and they will, in general depend on the drive v. We sometimes refer to them as *conditional* Lyapunov exponents. They serve as a generalized idea of eigenvalues for stability. When the spectrum of exponents is all negative, then part (iii) is satisfied and we have stability of the w system with respect to the drive v(t). This generalizes the usual uses of the fundamental stability theorem and gives us a new tool for building stable, chaotically driven systems.

3. SYNCHRONIZATION OF CHAOTICALLY DRIVEN SYSTEMS

The synchronization of two systems usually requires that, in some sense, two identical systems be built. We can use a constructive approach and consider not two isolated systems, but one whole system and one made from a duplicate of a *sub*system of the original [1,2]. If the subsystem is a stable subsystem in the above terms, we have synchronizable systems by driving the duplicate with signals from the original system that are absent in the duplicate. This is shown schematically in Fig. 2. If the original system has dimension n (=$m+l$) and the stable subsystem has dimension l, the dimension of the new compound system is $n+l$.

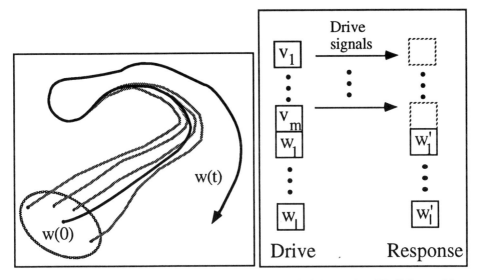

Fig. 1. Nearby trajectories converging to w(t).   Fig. 2. Schematic of Building a driven subsystem.

The concept of stable subsystems is rather new and non-trivial in nonlinear systems. Note that the Lyapunov exponent spectrum for a subsystem of a nonlinear system is generally not a subset of the Lyapunov spectrum for the whole system. Finding stable subsystems is therefore also a non-trivial problem. Testing the (conditional) Lyapunov exponents of the subsystems is the only way known for now. We comment more on this later.

The equations of motion are easily derived as follows. Suppose the drive system is divided into two sub systems with variables $v=(v_1,...,v_m)$ and $w=(w_1,...,w_l)$ with $n=m+l$ and with equations of motion $v=f(v,w)$ and $w=h(v,w)$. Then if the w equations have all negative Lyapunov exponents, this is a stable subsystem and a candidate for duplication. Our system of synchronized nonlinear subsystems is $\dot{v}=f(v,w)$, $\dot{w}=h(v,w)$, and $\dot{w}'=h(v,w')$, where w and w' will synchronize.

We can get an estimate of the convergence rates for synchronization, i.e. $\Delta w \rightarrow 0$. The estimate is essentially for small differences (Δw), but, in practice, appears to work well in many systems for differences on the order of the attractor size. This probably will depend strongly on the shape of the basin of attraction for the response system as well the the general flow of the system from the starting point to the true trajectory. Thus we urge caution in using the convergence rates for large $|\Delta w|$ values.

To estimate actual convergence rates we want to get an estimate of the average convergence rates and their associated directions. This translates into finding the (conditional) Lyapunov exponents and their eigenvectors. We calculate these from the principal matrix solution Z(t) for the Δw equation [3,4]: $\dot{Z}=D_w(w,v) Z$. The Lyapunov exponents are given by $\lambda_i = (1/t)\, ln\, v_i$ for large t, where v_i are the eigenvalues of Z(t), each associated with an eigenvector ζ_i. Calculation of these quantities is covered in ref [1].

We now approximate the convergence as $\Delta w(t) = e^{At} \Delta w(0)$, where A is generated by transforming the diagonal matrix with λ_i's on the diagonal back to the original coordinate system. We have shown that this can lead to a good approximation to the convergence, including effects associated with non-orthogonal eigenvectors[1].

We can ask further questions about the robustness of such synchronization, namely, if μ is the vector of parameters in the h(v,w) vector field, what is the effect of having slightly different values μ' in the response vector field? This is a situation one would expect in any application. If $\Delta w=w'-w$, where w and w' have different parameters, it is easy to show that $\dot{\Delta w}\Delta w=D_w h\, \Delta w + o(v,w)+h(v,u,\mu')-h(v,u,\mu)$. Note that the above theorem for uniform asymptotic stability does not apply. Because of the extra terms h(v,u,μ') and h(v,u,μ), criteria (i) of the theorem is not satisfied.

We can get an estimate of Δw by assuming that $\Delta w(0) \ll 1$. That is the systems have nearly identical initial conditions and we can ignore o(v,w) for times less than some t_1. Then, letting b_1 be an upper bound for $|h(v,u,\mu') - h(v,u,\mu)|$, c_2 be the largest Lyapunov exponent, and c_1 be a scale factor, it is straightforward to show [1]

$$|\Delta w| \leq \frac{c_1 b_1}{c_2}. \qquad (2)$$

So, if the differences in parameters are not large (which determines b_1), Δw will remain small and the systems will remain near synchrony.

In order to test all these ideas we performed numerical tests using this scheme with Lorenz, Rössler, and hysteretic Newcomb systems[1,2]. The integration was done with a Runge-Kutta 4-5 scheme with adaptable step size[3].

For the Lorenz system in the chaotic regime (σ=10, r=24.06 to > 200.0, and b=8/3) calculations of the conditional Lyapunov exponents for two-dimensional subsystems show that we can construct a stable subsystem by choosing either[2] w'=(x,z) or w'=(y,z). Fig. 3 shows the convergence of a time series for the y driven (x,z) response. Fig. 4 shows the same results for the trajectory of the (x,z) subsystem. This system appears to have only one basin of attraction. That is, all initial conditions for w' will eventually synchronize to the drive w values.

We tested the robustness against parameter variation by changing the parameter of the response by 10%. The results are shown in Fig. 5. Our estimate for $|\Delta w|$ agrees well with the numerical tests.

To test the practical application of these concepts we built an electronic circuit based on the hysteretic circuit of Newcomb[4]. This consisted of an unstable oscillator whose focus of instability could be switched from a positive fixed voltage to a negative one and back again, and a hysteretic subsystem which did the switching of the unstable focus based on the magnitude of one of the voltages in the unstable oscillator (when it got too large the hysteretic part would switch focii and cause the unstable oscillator to spiral out from another point). Details of this circuit are published elsewhere[5].

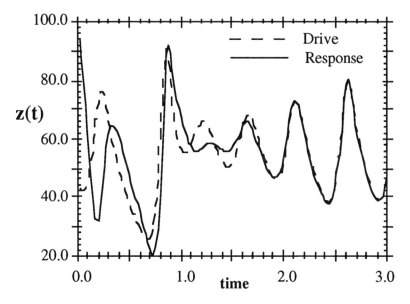

Fig. 3. Convergence of the z-component of the response to the drive z component.

This circuit had a stable subsystem consisting mostly of the unstable oscillator subcircuit. We constructed our response from a circuit with this subcircuit in it. The attractor is shown below along with that calculated from the model for the circuit.

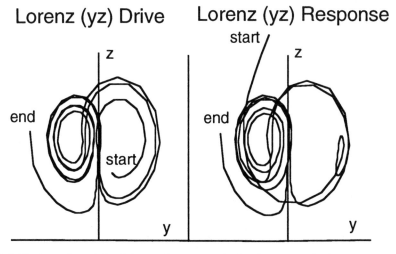

Fig. 4. Trajectories of the drive and response (yz) subsystems during convergence.

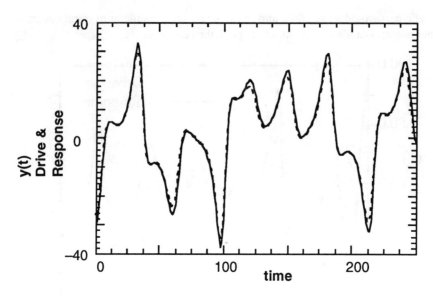

Fig. 5. y(t) for drive and response when response parameters are ~5% different from drive

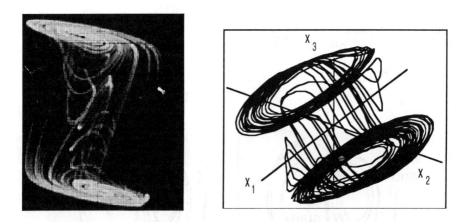

Fig. 6. Attractor for circuit (a) from oscilloscope and (b) as calculated.

An easy way to demonstrate synchronization in a pair of circuits is to display voltages on the oscilloscope from the same points the the stable subsystems, using one as the x and the other as the y scope drives. This is shown in Fig. 7. This figure also shows the effect of varying a resistor in the response circuit by the percentages shown. This displays the robustness shown by the theory and the numerical models actually exists in real circuits. We have built several other circuits which show the same capability of synchronization.

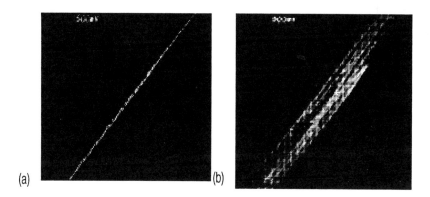

Fig. 7. voltages in drive vs. response: (a) same parameter, (b) 50% change in parameter.

4. CASCADING SYNCHRONIZED SYSTEMS

In some cases, such as the Lorenz equations, it may be possible to decompose the drive system into more than one stable subsystem. In this case, the individual response subsystems may be cascaded. The most interesting way to cascade the response subsystems is in a manner that causes the original drive signal to be reproduced. This is shown schematically in Fig. 8. One may take the difference between the original driving signal and its reproduction to produce a "chaos filter"; the output of this operation will be zero if the right chaotic signal is used for driving, but finite for any other signal. This is a type of signal recognition system. When the right subsystems are used, the response systems actually reproduce every signal in the drive system by receiving one signal. One may think of this system as a means for transforming the drive system attractor, which is a multidimensional pattern in phase space. When the response systems are synchronized to the drive system, the drive attractor is transformed into itself; otherwise, the drive attractor is transformed into some other pattern.

We have built a 4-dimensional chaotic circuit to test these ideas[6]. When we cascaded two response subsystems of this circuit, the output of the final response matched the initial drive signal to within 2%. In order to demonstrate the predictability of the pattern of behavior of the response circuit, we set up an experiment where a parameter in the response system was varied based on the value of the corresponding parameter in the drive system.

We based our control experiment on the behavior of return maps for the response system. A return map for a driven system is generated by strobing the some variable of the driven system when the drive crosses zero. The value of this variable at one drive crossing is plotted vs. the value of the variable at the next drive crossing. For the circuit used here, the average location of this return map shifts as the parameter difference between the driven and response systems changes. When a parameter in the drive system was varied, we used an analog circuit to find the average of the return map and another analog circuit to adjust the corresponding parameter in the response circuit. Figure 9 shows how the parameter in the response circuit tracked changes in the drive parameter.

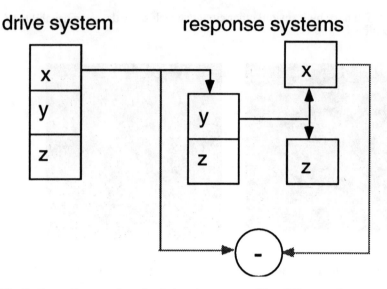

Fig. 8. Cascading synchronized chaotic systems. The difference between drive and response x may be taken to create a "chaos filter".

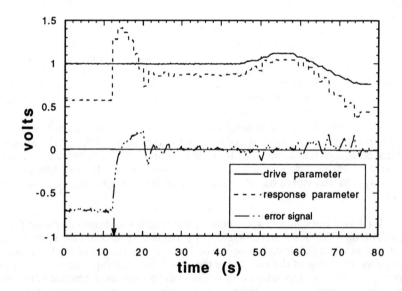

Fig. 9. Controlling a parameter in a chaotic response circuit by varying a parameter in a chaotic drive circuit. The control was turned on at the arrow.

5. CONCLUSIONS

The past decade was notable for progress in using the techniques of nonlinear dynamics to analyze physical systems' behavior. We now have the opportunity to move from passive analysis to a balanced combination of analysis and synthesis. This will be the only way the complicated nonlinear behavior we all study will move toward "usefulness". The two major themes involving use of concepts from nonlinear stability theory and constructive use of dynamical systems in this paper are an example of this approach.

This is not a matter of now applying nonlinear concepts to present engineering situations, but of actually being creative with those concepts to produce new systems and new concepts which use the unique properties of behavior such as chaos. These attempts actually enrich the entire field, including the theoretical/mathematical end. In the process of trying to find uses for chaos, unanswered and previously un-thought-of questions come up. Many of which are deep and fascinating. We list some related to our work below after mentioning potential applications.

In the realm of driven, synchronized chaotic systems several application possibilities come to mind. One now has a method by which remote systems can (as a whole) behave chaotically, but whose parts will remain stably synchronized. This suggests uses for communications, perhaps with chaotic signals serving as broad band carriers driving synchronized receivers which can strip away the chaos and reveal the information message. This in turn suggests that a more sophisticated version of the synchronization setup might be suitable for analog encryption or signal hiding[6,7]. Electronic versions of synchronized maps may be fitting for digital methods.

A broader vision of these results leads one to realize that when responses are stable to chaotic driving, they will always display the same pattern of behavior given the same driving signal. This leads to the possibility of designing nonlinear systems as signal identifiers and specialized filters.

Once one begins to use chaotic driving, deficiencies in the theory immediately show up. For one thing, until now, almost no one has consider chaotic drives. There is much to read about periodic driving, but almost nothing about chaotic driving. The use of a chaotic drive may be analyzable in terms of stochastic concepts, but the chaotic drive is not the same as a stochastic drive. For example, for a flow a chaotic drive is smooth and deterministic.

The mathematical questions that come up are deeply related to other concepts in nonlinear dynamics. How does one characterize the stable (or unstable) subsystems of a nonlinear system? It is doubtful that these are a topological invariant[1]. How are they related to the stable and unstable manifolds (as they ought to be)? Are there any apriori ways to find stable subsystems without actually doing a numerical calculation of the Lyapunov exponents? Is there a direct relation between the number of drive signals (say, m) and the number of positive Lyapunov exponents ($=m$?), as might be expected? The Rössler system shows that a chaotic system need not have any stable subsystems, but the question remains open as to whether a chaotic system can have all of its subsystems stable (the Lorenz system almost does for a large parameter range). The use of pseudo periodic driving shows that one can use a homotopic approach to varying the drive from a periodic one to a chaotic one with the possibility that we may be able to go all the way to a pure chaotic signal and keep the response stable and the motion topologically similar to the original periodically driven case. When can this be done? What bifurcations or crises are possible along the way? What happens to the basins of attraction in higher dimensions? Will fractal basins remain fractal in a pseudoperiodically driven system?

Certainly, we believe that the research presented here is merely scratching the surface of applications using chaotic driving. One can now design all sorts of interconnected systems made from autonomous systems or subsystems. Others have also begun to find unusual applications for chaotic systems[8-11]. The number of permutations are impossible for any one researcher to investigate and the number of different nonlinear behaviors must surely be likewise.

6. REFERENCES

1. Louis M. Pecora and Thomas L. Carroll, Phys. Rev. A44, 2374 (1991).
2. Louis M. Pecora and Thomas L. Carroll, Phys. Rev. Lett. 64, 821 (1990).
3. W.H. Press, B.P. Flannery, S.A. Teukolsky, W.T. Vetterling, *Numerical Recipes* (Cambridge Univ. Press, New York, 1990).
4. R.W. Newcomb and S. Sathyan, IEEE Trans. Circuits Syst. 30, 54 (1983).
5. Thomas L. Carroll and Louis M. Pecora, IEEE Trans. Circuits Syst. 38, 453 (1991).
6. T. L. Carroll and L. M. Pecora, Int. J. Bifurcations and Chaos 2, 659 (1992).
7. T. L. Carroll and L. M. Pecora, Physica D, in press.
8. J. Gullicksen, R. Sherman, M. Steinberg, J. Huang, M. de Sousa Vieira, M. A. Lieberman, W. Wonchoba, and P. Khoury, proceedings 1st Experimental Chaos Conf. (World Scientific, 1992).
9. M. Spano and W. Ditto, proceedings 1st Experimental Chaos Conf. (World Scientific, 1992).
10. E. R. Hunt, proceedings 1st Experimental Chaos Conf. (World Scientific, 1992).
11. T. Shinbrot, E. Ott, C. Grebogi and J. A. Yorke, Phys. Rev. Lett. 65, 3215 (1990).

Controlling Chaos

Mark L. Spano
Naval Surface Warfare Center

and

William L. Ditto
Georgia Institute of Technology

Abstract

The concepts of chaos and its control are reviewed. Both are discussed from an experimental as well as a theoretical viewpoint. Examples are then given of the control of chaos in a diverse set of experimental systems. Current and future applications are discussed.

Introduction

Chaos - A rough, unordered mass of things.

Ovid, *"Metamorphoses"*

Shakespeare once asked, "What's in a name?" The answer is nothing and everything. Nothing because "A rose by any other name would smell as sweet." And yet, without a name Shakespeare would not have been able to write about that rose or to distinguish it from other flowers that smell less pleasant. So also with chaos. The dynamical phenomenon we call chaos has always existed, but until its naming we had no way of distinguishing it from other aspects of nature such as randomness or order. And lacking any way to treat it as a separate entity, we would not even be able to conceive of manipulating it or controlling it. Thus the naming of chaos by James Yorke[1] in 1975 was an essential first step towards identifying it and learning to control it.

From this identification came the recognition that chaos is pervasive in our world. Chaos was first suspected in the weather patterns[2] that govern our lives. It is easy to make simple electronic circuits that are chaotic.[3] Mechanical systems on such wildly different scales as laboratory pendula[4] and orbiting planets[5] have been shown to exhibit chaos. Laser emissions[6] can fluctuate chaotically. The human heart shows evidence[7] of beating to chaotic rhythms. Chemical reactions[8,9] oscillate in chaotic fashion. Of these diverse systems, we have learned to control all of those that are on the smaller scale. Systems on a more universal scale such as the weather and the planets remain beyond our control.

In what follows we will discuss the concept of chaos, from both a theoretical and a laboratory viewpoint, always with an eye towards methods of controlling and exploiting it. Afterwards we will look at the application of these methods to the systems mentioned above.

© 1994 American Institute of Physics

The Concept of Chaos

A violent order is disorder; and
A great disorder is an order. These
Two things are one.

Wallace Stevens, *"Connoisseur of Chaos"*

Before its discovery and naming, chaos was inevitably confused with randomness and indeterminacy. Because many systems *appeared* random, they were actually thought to *be* random. This was true despite the fact that many of these systems seemed to fall into periods of almost periodic behavior before returning to more "random" motion. Indeed this observation leads to one of the definitions of chaos: the superposition of a very large number of periodic motions. A chaotic system may dwell for a brief time on a motion that is very nearly periodic and then may change to another motion that is periodic with a period that is perhaps five times that of the original motion and so on. This constant evolution from one (unstable) periodic motion to another produces a long-term impression of randomness, while showing short-term glimpses of order. These glimpses are not misleading, since chaos is deterministic and not random in nature.

A second definition of chaos is this: the sensitivity of a system to small changes in its initial conditions. This is the "butterfly effect" of legend. The story goes as follows: the small breeze created by a butterfly's wings in China may eventually mean the difference between a sunny day and a torrential rainstorm in California. If you visualize the motion of a chaotic system in phase space (the space composed of the usual spatial dimensions plus their accompanying momentum dimensions), you soon see that all these unstable periodic motions have corresponding trajectories in phase space. And since there are a very large number of these unstable motions in a bounded space, the trajectories corresponding to each unstable periodic motion must be packed quite closely together. Thus only a very small perturbation may push the system from one unstable periodic motion to any of a large number of others.

This last is both the hope and the despair of those who have to deal with chaotic systems. It is the despair because it effectively renders long term prediction of these systems impossible. For instance, in a computer simulation a small perturbation may be introduced by numerical round-off, while, in the real world, systems are invariably perturbed by noise. Thus, if a system is chaotic, these small perturbations quickly (indeed, exponentially) grow until they completely change the behavior of the system.

Paradoxically, the cause of the despair is also the reason to hope. Because if a system is so sensitive to small changes, could not small changes be used to control it? This realization led Ott, Grebogi and Yorke[10] (OGY) to propose an ingenious and versatile method for the control of chaos, the details of which are given in the following section.

The Idea of Control

At which the universal host up sent
A shout that tore hell's concave, and beyond
Frighted the reign of Chaos and old Night.

John Milton, "Paradise Lost"

The starting point for the control of chaos is the phase space of the system. Dynamical systems trace smooth trajectories in phase space as they proceed forward in time. For periodic systems these trajectories are curves that are traced repeatedly every period of the system. For example, the phase space of a driven simple pendulum consists of one dimension for its position and one dimension for its momentum. The trajectory of the pendulum in this space is simply a circle which is traced over and over, as in Fig. 1a. This trajectory contains all the information that is needed to predict the future dynamics of the system. In fact it contains too much information. A more useful representation can be obtained by cutting through this phase space with a plane which intersects the circle in two places. The infinite number of points on the circle trajectory has been reduced to merely two. If we further confine ourselves to *directed* piercings of the plane, we are left with only a single point (Fig. 1b). Such a *Poincaré section* reduces our information to a manageable level.

If our experiment were more complex, perhaps a driven pendulum with two masses placed at different points on a flexible string or the simple pendulum driven well beyond its linear regime, more complex motions might occur. One such might be a motion in which the trajectory traced a double loop in its phase space (Fig. 1c). A Poincaré section through this trajectory would show two points (Fig. 1d). This pendulum could, of course, still execute a simple, single loop trajectory given the right initial conditions. This single loop defines the *period one* trajectory of the system. The double loop trajectory takes twice as long and is accordingly dubbed *period two*. The usefulness of the Poincaré section is now apparent: the number of points in the section immediately reveals the underlying periodicity of the system. Generally, any periodic system of period n will have a section which consists of a finite set of n distinct points which reflect the fundamental periodicity of the system.

By way of contrast, a truly random system will behave in such a way that the trajectories in phase space wander over the entire volume of space available to the system. Its corresponding section will be filled uniformly and densely.

Chaos falls between these two extremes. Since chaos is a superposition of a number of periodic motions, one might expect to see a finite number of points indicating several periodic motions characteristic of the chaotic section. This is true, as far as it goes. However, since chaos is the superposition of a large (read *infinite*) number of periodic motions, the number of points in the section is also infinite. In general, these points form an extended geometric structure, called the system's *chaotic attractor*, which is not a finite set of points (*i.e.*, does not represent periodic motion) and which also does not fill space (*i.e.*, is not random), as shown in Fig. 1f. (Generally a chaotic attractor is a fractal object.) Knowledge of this attractor and of its response to small perturbations of the system are the only ingredients that are necessary for the control of chaos.

In order to control chaos according to the OGY scheme, it is only necessary to identify an unstable periodic point in the attractor, to characterize the shape of the attractor locally around that point and to determine the response of the attractor at that point to an external stimulus. Let's look at each of these three steps in detail.

The identification of an unstable periodic point is easy since every point in the attractor represents an unstable periodic motion with some period n. (Of course it is easier to find points with low order periodicity (small n), but chaos control has been successfully implemented to select periodic motions of order up to about 30!) This is one of the strengths of the OGY method of control: the ability to control on any periodic motion from the infinity of motions present in the system. This capability would allow an engineer to design highly flexible systems employing chaos control.

Characterizing the shape of the attractor is also easy. Once we have determined which unstable fixed point we wish to control about, we observe the motion of the point representing the current state of the system (*system state point*) on the attractor. In low dimensional chaotic systems, this point will occasionally approach the vicinity of our chosen unstable fixed point and then move away again. It turns out that, in the neighborhood of the unstable fixed point, the approach is consistently along the same direction (called the *stable direction*) and the departure along another direction (called the *unstable direction*), as indicated in Fig. 2. These two directions, one of which is stable (incoming) and the other of which is unstable (departing), form a saddle around the unstable fixed point. These *eigenvectors*[11], along with the speed with which the system state point approaches or departs the vicinity of the unstable fixed point (*stable* or *unstable eigenvalues* respectively), are all that is necessary to characterize the shape of the attractor locally around our chosen point.

The final criterion, determining the response of the attractor to an external stimulus, is perhaps the most difficult of the three to learn. Here we must introduce a change into the system by varying one of the system parameters. The difficulty lies in first identifying the system parameters and then locating one that may be quickly changed. (More on this will be discussed in the experimental section below.) Once such an identification has been made, the position of the unstable fixed point is measured for several values of the parameter just slightly different from the nominal value (Fig. 3). These measurements tell us the change of the attractor position with respect to a change in the system parameter and complete the knowledge of the system needed for control.

All of this information is fed into the OGY control equations:

$$\delta p = C(\xi_n - \xi_f) \cdot \mathbf{f}_u \tag{1}$$

where

$$C = \frac{\lambda_u}{\lambda_u - 1} \frac{1}{\mathbf{g} \cdot \mathbf{f}_u}. \tag{2}$$

δp is the amount one needs to change the system parameter from its nominal value in order to control the system. The perturbation δp is updated once every period of the system. It is proportional to the distance of the system state point, ξ_n, from the fixed point, ξ_f, projected onto the unstable direction, f_u. The constant C is itself calculable from the change of the attractor position with respect to a change in the system parameter, g, projected onto the unstable direction and from the value of the unstable eigenvalue, λ_u. All of these are known from our earlier measurements.

An example may serve to clarify the matter. Fig. 4a shows an unstable fixed point with its stable and unstable directions. The system state point, ξ_n, is shown approaching the fixed point near the stable *manifold*. If left to itself, the natural motion of the system would be to approach the unstable fixed point along the stable direction and then move away from the fixed point along the unstable manifold (towards the upper right). However, if we introduce the OGY perturbation into the system, we move the unstable fixed point and its accompanying manifolds to the other side of the system state point (Fig. 4b). The new motion of the system state point will still be along the stable manifold with motion away from the *new* fixed point along the unstable manifold (towards the lower left). However this is motion towards the old unstable fixed point. Thus the OGY scheme has controlled the system exactly onto the unstable fixed point (Fig. 4c). In a perfect world, it would remain there forever. However if there is even the slightest error in the calculation of C, it will only move close to the fixed point. Or in the presence of noise it will be perturbed from the fixed point after it has reached it. Both of these problems are easily handled by repeating the OGY control once each period of the system.

This act of moving the saddle point to control the motion of the system state point is similar to balancing a baseball on a saddle or a baseball bat on the palm of your hand. The former case is directly analogous, with one stable and one unstable direction. Small movements of the saddle keep the ball at the point of unstable equilibrium; it is not necessary to change the shape or nature of the saddle itself.

Many methods of "control" have been proposed. Those related to the OGY method[9,12] use perturbations that are typically small (a few percent of the magnitude of the accessible system parameter). Thus it is easy to design systems that implement this control. This is in sharp contrast to the model-based methods[13] which seek to fundamentally change the nature of the chaotic system under study by introducing "perturbations" to the adjustable parameter that are comparable to its size. Such methods are more properly named the "removal of chaos" rather than "control" because they fundamentally change the nature of the system under study. The gentle nudges prescribed by OGY control leave the fundamental dynamics of the system essentially unchanged. The fixed point and its manifolds are only shifted slightly by the perturbation and, between perturbations, the motion of the system remains chaotic, moving on an attractor that differs little from the original. It is only in the long term, when *averaged* over many periods and over motion on several slightly different attractors, that the motion is non-chaotic. In this sense the OGY method is truly a *control* of chaos and hence it is able to select any of the periodic orbits that are embedded in the chaotic motion.

Experimental Realities

In all chaos there is a cosmos, in all disorder a secret order.

Carl Jung

An experimental physicist's first reaction to the OGY scheme might well be one of horror; in any real-world system the number of degrees of freedom (positions plus momenta) may be quite large, possibly even infinite and the OGY scheme requires the measurement of all of them! Two facts conspire to make the OGY method not only possible but also simple to implement.

The first is that many physical systems, while technically infinite dimensional, actually only express a small number of these dimensions. Thus even though the magnetomechanical ribbon (discussed below) is a distributed mass system and therefore technically infinite dimensional, its motion is quite low dimensional and its attractor is well represented in a three dimensional phase space and hence a two dimensional Poincaré section. Thus only two experimental variables need to be measured to correctly characterize its chaotic attractor.

The second fact of importance to the experimentalist is that the Poincaré section may be replaced in the OGY scheme by a *delay coordinate embedding*. This embedding relies on the intuition that the information obtained by measuring all the system variables (positions and momenta) at one given time may also be obtained by measuring only one system variable at several subsequent times. The system state vector (the set of all the positions and momenta, $\xi_n = (x^1(t_n), x^2(t_n), \ldots p^1(t_n), p^2(t_n), \ldots))$, is replaced by a delay coordinate vector, $(x^i(t_n), x^i(t_n - \Delta t), x^i(t_n - 2\Delta t), x^i(t_n - 3\Delta t), \ldots)$, where the superscript i indicates one particular experimental measurable and for some appropriately chosen delay, Δt. (If the system is periodically driven, Δt is usually chosen to be the period of the driving force. In non-driven systems the appropriate delay is usually related to some fundamental time scale inherent in the system.) In fact it may be shown that the attractor as represented in the phase space and the attractor reconstructed from the delay coordinate embedding have the same topological properties, thereby allowing one to be substituted for the other with (relative) impunity.[14]

With these two simplifications in hand, control of low dimensional chaotic systems becomes simple. In fact it turns out that the OGY method is also extremely robust. By this we mean that, as long as the perturbation δp is sufficient to move the saddle point to the opposite side of the system state point as shown in Fig. 4b, it does not matter if the saddle is moved well past the state point or just barely past it. The corollary to this is that the control scheme is quite resistant to external noise and to errors arising from imprecision in the measurement of the position and shape of the attractor.

We note in passing that eqs. 1 and 2 give rise to two variant methods of control. In the first method, one completely measures all the items mentioned above, calculates the constant C, and then proceeds to control the system. The alternative arises from the fact that, although many values (λ_u, f_u, and g) are needed to characterize the attractor, they are all folded into the single constant C. Thus, if a period of adjustment can be tolerated, one might simply guess a value of C and then adjust it empirically until control is achieved. This latter method, requiring knowledge only of the location of the desired fixed point and an educated guess at C makes this quite suitable for incorporation into automatic control systems.

Implementations of Control

*Not chaos-like together crushed and bruised,
But, as the world, harmoniously confused*

Alexander Pope, "Windsor Forest"

Mechanics

The first experimental control of chaos was achieved in a magnetoelastic ribbon experiment[4]. This system is sketched in Fig. 5. The ribbon, an amorphous magnetic material about 100 mm long by 3 mm wide by 0.025 mm thick, is extremely nonlinear in its response to a magnetic field. In zero field it is stiff enough to support its own weight in an upright position. When a small magnetic field is applied, its stiffness decreases by an order of magnitude and the ribbon buckles under its own weight. At still larger fields, the material again becomes stiff and resumes an upright position. If a magnetic field, H_{dc}, is applied to start the ribbon in the soft regime and to that is added an ac magnetic field, H_{ac}, that alternately takes the ribbon from soft to stiff and back again at some frequency, f, the ribbon will oscillate about its equilibrium point. At high enough values of H_{ac} and f, the ribbon's motion will become chaotic.

The position of the ribbon is measured once each period, $T = 1/f \approx 1Hz$, of the driving field by means of a beam of light directed at a small spot on the ribbon. The intensity of the reflected light varies as $1/r^2$, yielding highly accurate and noise-free measurements of the ribbon position. These are sent to the computer controlling the experiment, which also control the strengths of the magnetic fields. In theory, either of the three system parameters, H_{dc}, H_{ac} or f, could be used to control the chaos. We chose to use H_{dc}. The gray data in Fig. 6a show the position of the ribbon as a function of time when the ribbon is moving chaotically. When the computer initiates the period 1 control, the position of the ribbon (measured once each period T; i.e., stroboscopically) is shown in black in Fig. 6a. The corresponding attractor (plotted as a delay coordinate embedding and with the same color coding) is shown in Fig. 6b.

As mentioned above the OGY method may select any of a number of periods from the chaotic signal. Fig. 7 shows the same system switching from chaos to period 4 to period 1 to period 2 to period 1 and finally back to period 4. There are interludes of chaos between each section of control because the computer must wait for the ribbon's position to approach the vicinity of the new unstable fixed point. This time can actually be decreased by a new technique called *targeting*, [15,16], in which the sensitivity of the chaotic system to small perturbations plus a *global* (but still empirical) knowledge of the chaotic attractor are employed to direct the system rapidly to any desired point on the attractor.

The question arises as to whether one might control about a point that does not lie on the chaotic attractor. It turns out that the difficulty in such a case lies not in the control itself, but in getting the system to the vicinity of the desired point. For if the point in question is not on the attractor, it will never be visited. However, once control is established at a "normal" unstable fixed point, it is often possible to adjust some system parameter (in this case the temperature) so that the attractor smoothly changes to one that does not include the point in question. Yet it has been shown experimentally that the control will be maintained about this point. The reason is that, although the point is no longer part of the attractor, the local geometry about the point has not changed appreciably and thus the algorithm still calculates the proper control perturbations. This is an important result for understanding the robustness of the method even under large changes to the system itself.

Electronics

In an experiment[3] of great potential importance for our electronics-based world, Earle Hunt of Ohio University also used the empirical version of the OGY scheme to control the chaotic oscillations of a diode resonator. The significance of his experiment lies in the extension of the applicability of the method to high frequencies (53 kHz) and in his ability to select from the chaotic attractor periods up to period 23.

Lasers

In conjunction with Hunt, Rajarshi Roy et al.[6] of Georgia Tech controlled the chaotic oscillations of a chaotic multimode laser. This laser would autonomously become chaotic at pumping levels exceeding a certain threshold. The system is somewhat higher dimensional than the previous systems, making the control more difficult to obtain. Additionally the system is not periodically driven, making it necessary to choose a "natural periodicity characteristic of the system" as the time scale for analyzing their data and applying the control signals. A third intricacy is that the natural frequency mentioned above is on the order of hundreds of kilohertz. They developed a method related to the OGY method to control of the system up to period 9. Thus the laser could be operated stably at power levels exceeding by as much as a factor of 15 those which might be attained without chaos control.

Biology

The authors have recently extended the techniques of chaos control to the field of biology.[7] In this experiment a section of a rabbit's heart was induced to beat chaotically by the injection of the drug ouabain. The measurable quantity here is the interval between heartbeats, which is about 0.8 sec in the healthy tissue. The effect of the drug is to accelerate the heartbeat and to cause the interbeat intervals to vary chaotically. The chaotic attractor for this situation is shown in Fig. 8a.

There are several factors that distinguish this experiment from the earlier successes at chaos control. The first is that the data for the delay coordinate embedding are no longer equally spaced in time, but are rather intervals between events. The second factor is that there are no adjustable system parameters that may be modified quickly enough to implement the unmodified OGY control scheme. The authors accordingly developed a method whereby the perturbation is applied directly to the system variables, rather than to some system parameter. This method uses the same local linearization around the unstable fixed point as used by OGY, but instead of moving the saddle point to the system state point, this method moves the system state point back to the saddle by giving the state point small nudges equal in magnitude but opposite in sign to those yielded by eqs. 1 and 2. To calculate these nudges it was necessary to use the observed chaotic attractor to predict the time of the next heartbeat. Then an electrical stimulus was used to induce a heartbeat before the predicted natural beat could occur, thereby shortening the next interbeat interval. The amount of time to advance the next beat was calculated so as to place the system state point (Fig. 8b) directly onto the stable direction of the attractor. Thus the succeeding beat would tend to naturally move toward the desired unstable fixed point rather than away from it.

The results are shown in Fig. 9. It was not possible in these early experiments to achieve a good period 1 (i.e., "normal") heartbeat. But it was possible to control the chaos consistently into a period 3 beat, which, while not optimal, is better for pumping blood than chaotic beating. As for the relevance to human heart arrhythmias, there is evidence that atrial and ventricular fibrillation may be examples of chaos. Thus future work may be able to present strategies for dealing with these serious and widespread arrhythmias.

Chemistry

The best known oscillatory chemical reaction, the Belousov-Zhabotinsky reaction, may exhibit either periodic or chaotic signatures, depending on the flow rates of the reactants into the tank. These rates comprise the system parameters to be used for control. The variable that is measured is the voltage of a bromide electrode inserted into the tank. Using the OGY method, Petrov et al.[8] at West Virginia University were able to stabilize both period 1 and period 2 oscillations in their reactor.

Rollins et al.[9] at Ohio University have also applied chaos control to a chemical system. They developed a recursive method (based on the Dressler and Nitsche modification of the OGY method) that makes available a wider choice of accessible system parameter for use in controlling the chaos.

Future of Control

Chaos often breeds life, when order breeds habit.
Henry B. Adams, "The Education of Henry Adams"

Theory

Two fundamental questions control future chaos control theories. The first is the problem of controlling higher dimensional chaos in physical systems. A start has been made in this direction in a recent paper by Auerbach, Grebogi, Ott and Yorke[17] on controlling chaos in systems of arbitrarily high dimensions. The method can be "implemented directly from time series data, irrespective of the overall dimension of the phase space." Although applied in numerical studies, this method has yet to be tested experimentally.

The second question has yet to be addressed: the problem of control in a spatio-temporal system. Many fluid dynamical and biological systems may be modeled as a multi-dimensional array of oscillators connected to each in some complex fashion. Such systems exhibit both spatial as well as temporal chaos. The study of such systems, while of obvious importance for real-world applications, is as yet in its infancy.

Experiment

The importance of chaos control in electronic circuits was mentioned above and should be stressed once again. In a world dominated by electronics, the ability not only to remove chaos where it is not wanted, but also to make more flexible circuitry by exploiting chaos and its control could have a tremendous impact on our lives. The control of chemical reactions is no less important, since an increase of only a few percent in the efficiency of a commercial chemical reaction could conceivably save millions of dollars. The 15-fold increase in the power at which a laser can be operated stably promises new applications for these modern tools. And the ability to control one of the great diseases of modern times could be a boon beyond reckoning. The challenge for experimentalists now lies not only in achieving control of higher dimensional and spatio-temporal chaos, but also in guiding the achievements of the past three years out of the laboratory and into our lives.

References

[1] James Gleick, Chaos (Viking, New York, 1987), p. 65; see also T.-Y. Li and J. A. Yorke, *Amer. Math. Monthly* **82**, 985 (1975).

[2] E. N. Lorenz, *J. Atmos. Sci.* **20**, 130 (1963).

[3] E. R. Hunt, *Phys. Rev. Lett.* **67**, 53 (1991).

[4] W. L. Ditto. S. N. Rauseo and M. L. Spano, *Phys. Rev. Lett.* **65**, 3211 (1990).

[5] Gerald Jay Sussman and Jack Wisdom, *Science* **241**, 433 (1988).

[6] Rajarshi Roy, T. W. Murphy, Jr., T. D. Maier and Z. Gills, *Phys. Rev. Lett.* **68**, 1259 (1992).

[7] Alan Garfinkel, Mark L. Spano, William L. Ditto and James N. Weiss, *Science* **257**, 1230 (1992).

[8] Valery Petrov, Vilmos Gaspar, Jonathan Masere and Kenneth Showalter, *Nature* **361**, 240 (1993).

[9] R. W. Rollins, P. Parmananda and P. Sherard, *Phys. Rev. E* **47**, R780 (1993).

[10] E. Ott, C. Grebogi and J. A. Yorke, *Phys. Rev. Lett.* **64**, 1196 (1990).

[11] Actually these are the covariant eigenvectors, e_s and e_u, from which must be derived the contravariant eigenvectors, f_s and f_u. This is because the two directions are, in general, not orthogonal. See Ref. 9 for more detail.

[12] B. Peng, V. Petrov and K. Showalter, *J. Chem. Phys.* **95**, 4957 (1991).

[13] A. Hubler, *Helv. Phys. Acta* **62**, 291 (1989).

[14] However, see Ute Dressler and Gregor Nitsche, *Phys. Rev. Lett.* **68**, 1 (1992).

[15] T. Shinbrot, E. Ott, C. Grebogi and J. A. Yorke, *Phys. Rev. Lett.* **65**, 3215 (1990); T. Shinbrot, E. Ott, C. Grebogi and J. A. Yorke, *Phys. Rev. A* **45**, 4165 (1992).

[16] Troy Shinbrot, William Ditto, Celso Grebogi, Edward Ott, Mark Spano and James A. Yorke, *Phys. Rev. Lett.* **68**, 2863 (1992).

[17] Ditza Auerbach, Celso Grebogi, Edward Ott and James A. Yorke, *Phys. Rev. Lett.* **69**, 3479 (1992).

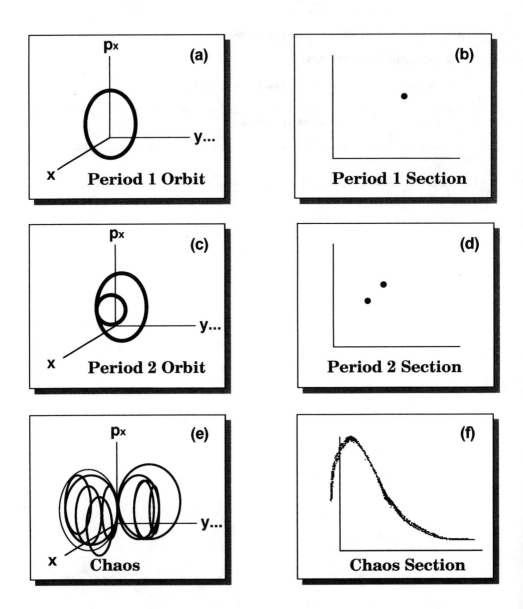

Fig. 1 Phase space trajectories (a, c, e) and their corresponding Poincaré sections (b, d, f) for period 1, period 2 and chaotic orbits.

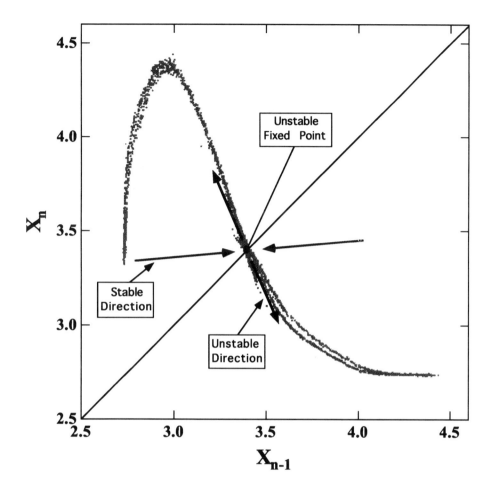

Fig. 2 The geometry around an unstable fixed point on a chaotic attractor. The shape is that of a saddle.

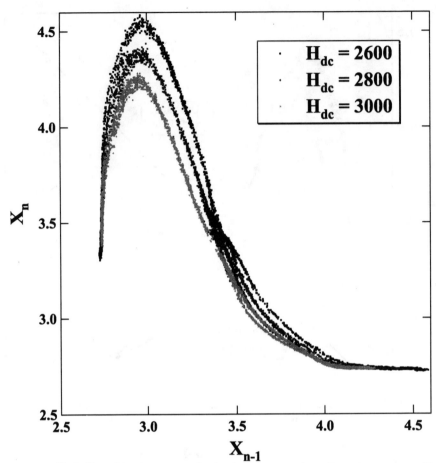

Fig. 3 Determining the response of a chaotic attractor to a change in a system parameter, H_{dc}. The change in the unstable fixed point position with a change in the parameter yields the vector quantity **g**.

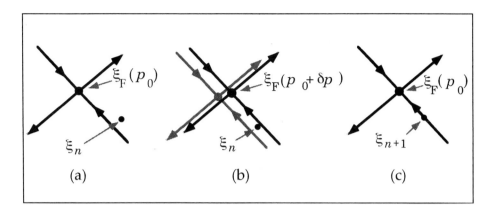

Fig. 4 An outline of the OGY control method: (a) the system state point approaches the unstable fixed point. It will naturally tend to move in along the stable direction and out along the unstable direction; (b) the saddle point (unstable fixed point) is positioned so that the system state point tends to move back toward the original unstable fixed point; (c) the system state point reaches the original unstable fixed point and the perturbation is turned off. In real-world systems with noise, the control would have to applied continuously to balance the state point on the saddle.

152 Controlling Chaos

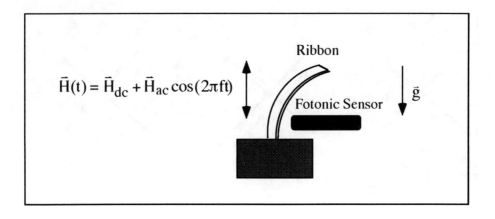

Fig. 5 Schematic of the magnetoelastic ribbon experiment.

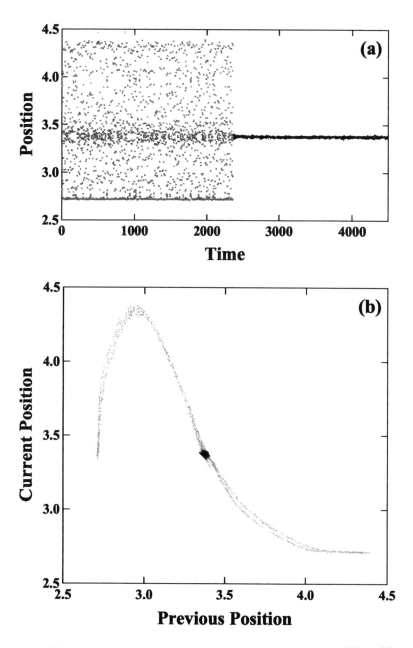

Fig. 6 (a) Ribbon position as a function of time for chaotic motion (gray) followed by controlled motion (black); (b) The same data presented as a delay coordinate embedding.

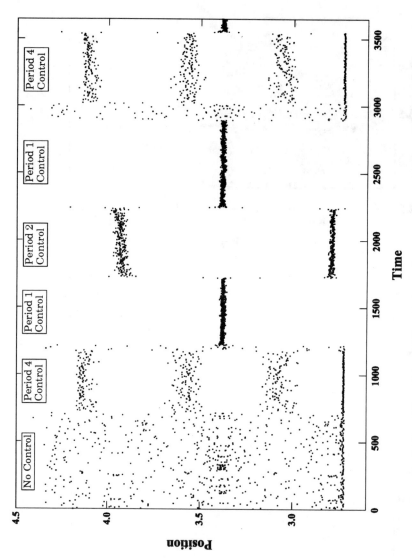

Fig. 7 The ability of the OGY method to select out various periodic motions from those comprising the chaotic motion. (Note that the fact that the motions presented here are all a power of two is purely coincidental; the control could have as easily been performed about the period 3 or period 5 motion instead.)

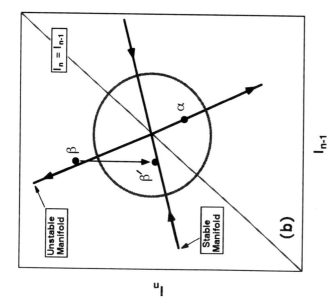
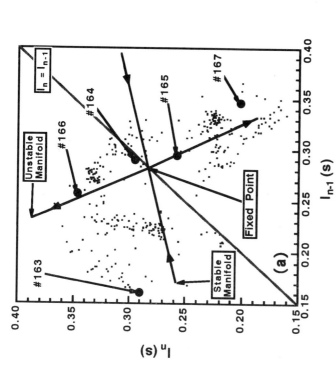

Fig. 8 Current interbeat interval, I_n, versus previous interbeat interval, I_{n-1}, for the rabbit heart experiment. (a) Starting with point 163, the system approaches the unstable fixed point along the stable direction and then leaves it along the unstable directions. (The unstable eigenvalue is negative, thereby causing subsequent point to alternate above and below the unstable fixed point. However the *distance* from the fixed point increases in an exponential fashion, as is typical of chaotic motion.) (b) An idealized drawing of the control. The point α should be identified with point 165 in (a). The next point would naturally tend to be β (point 166). However a stimulus is applied to the heart causing it to beat early and thereby shortening the next interbeat interval to β' on the stable manifold. Thus the next interval will tend to move along the stable manifold toward the fixed point.

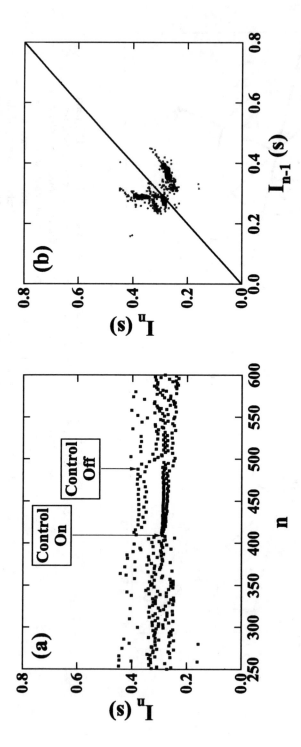

Fig. 9 (a) Interbeat intervals as a function of time for the rabbit heart experiment. (b) Delay coordinate embedding for the same data. The pre- and post-control data is gray, while the controlled data is shown in black.

SIGNAL MODELING
AND NOISE REDUCTION STRATEGIES

ADAPTIVE NONLINEAR DYNAMICAL PROCESSING FOR TIME SERIES ANALYSIS

J. S. Brush
J. B. Kadtke[*]
RTA Corporation, P.O. Box 5267, Springfield, VA 22150

ABSTRACT

Adaptive, or time-varying, modelling approaches to signal processing have typically been employed when non-stationarity is presumed to exist. In a nonlinear dynamics framework, the adaptive paradigm has an additional application in modelling which simultaneously addresses some of the limitations of local linear and static global methods, even for stationary situations. These new methods therefore have the ability to account for non-stationarity as well as nonlinear signal properties. We discuss the implementation of two continuous model update schemes, as well as applications to system characterization, parameter tracking, and transient detection in noise.

INTRODUCTION

One significant goal in the field of signal processing is the reduction of noise or other corrupting components to increase the relative strength of a particular signal of interest (SOI). The most straightforward and perhaps the most common method for accomplishing this is by filtering an original signal. This filtering is accomplished using linear time or frequency domain filters that reduce the frequency components of the original signal which are known to lie outside of the spectrum of the SOI. Such an approach, widely used in signal processing applications, makes several assumptions about the signals involved. One crucial assumption is that the signals are sufficiently linear so that the Fourier decomposition used to generate the power spectrum captures the essential signal information. Another important assumption is that enough of the signal characteristics are known beforehand to choose the unrelated parts of the spectrum to suppress via the filter. While some of these assumptions can be overcome by various techniques when the signal data has been previously obtained, real time applications of direct filtering methods can prove much more challenging.

---

[*] also at: Institute for Pure and Applied Physical Sciences (IPAPS), MS-Q0075, University of California at San Diego, La Jolla, CA 92093

One of the most important applications of filtering approaches is the concept of noise cancellation, which was proposed as early as 1967.[1] This idea constituted a fundamental advancement in the way noise reduction and signal separation could be performed, and was specifically amenable to real-time processing. In its most basic form, the (adaptive) noise canceler works by using a separate signal (called the reference signal) which is a real-time representation of the unwanted signal components, usually recorded by a separate sensor. The reference signal is continuously modified by a linear filter, whose output is subtracted from the original signal (the primary input) with the aim of leaving behind the SOI. The important concept behind this technique is that the coefficients of the filter are continuously adapted to be a best fit to the unwanted signal components, by requiring that the total signal power of the primary signal plus filter output is minimized. Adaptation of the filter means that the noise canceler can reduce unwanted components even for time-varying signals, because the spectral characteristics of the filter are not static. Also, using a separate realization of the unwanted signal component (the reference signal) which is actively subtracted from the primary signal means that signal separation can be performed even for signals whose frequency components overlap in the spectral domain.[2]

Although the noise cancellation scheme described above has been used successfully in many areas, the method begins to fail when applied to problems of noise reduction and signal separation for signals which are not well approximated by linear filters -- that is, for signals which are generated by processes which have significant nonlinearities or purely stochastic components. This point is of particular importance due to the realization in the last decade that many real-world systems are examples of 'chaotic' processes. These nonlinear processes often have simple dynamical descriptions whose behavior nevertheless is so complex as to appear random to most classical statistical tests. Although considerable work has been done on simple nonlinear adaptive filters, with significant improvements in performance for moderately nonlinear signals, the possibility exists for the formulation of adaptive nonlinear filters which exploit the mathematical and numerical tools developed for the nonlinear analysis of chaotic systems. During the past fifteen years, a large number of such tools have been developed for the analysis of chaotic systems, and the field has grown enormously. 'Chaotic adaptive filters', such as we propose here, may provide the ability to perform adaptive noise canceling even for signals which would normally require very high-order linear filters to model the reference signal. Thus, these filters could greatly improve performance in many real-world applications where conventional adaptive filters perform poorly.

We will show how a nonlinear adaptive filter can be constructed in a straightforward way using an architecture very close to conventional adaptive filter schemes. We demonstrate that a numerical implementation of the filter converges to the correct generating representation of a chaotic reference signal in an understandable way, and that the filter can adaptively track changes in the parameters of the generating process. We also describe the recovery of SOIs from various levels of obscuring noise, and we will discuss the application of these ideas to real-world data sets, where the character of the SOI and the obscuring signals are arbitrary.

TRADITIONAL LINEAR FILTERING

The theoretical framework for adaptive linear filters was first developed to a significant degree in the mid- and late 1960s, although a good deal of the fundamental work had been formulated as far back as 1948.[3,4] Much of the early results and applications are summarized in the excellent paper by Widrow, et al.[2] In its most basic form, an adaptive linear filter can be described as follows: consider a discretely sampled signal $x[i]$, which we assume to consist of a component of interest $s[i]$ and some additive corrupting signal or noise, $n[i]$, which may be of quite large relative amplitude. Thus,

$$x[i] = s[i] + n[i]. \qquad (1)$$

The signal $x[i]$ is termed the primary input, and we would like to construct a scheme which removes as much of $n[i]$ as possible. The approach is to separately measure another time series $\tilde{n}[i]$, called the reference signal, which is measured in such a way as to be as close to a pure representation of $n[i]$ as possible. In general there will always be some difference between the reference signal $\tilde{n}[i]$ and the actual component $n[i]$ due to differences in the measuring devices, relative locations of the sensors, and other corrupting noise components, so we attempt to model these differences (i.e. the transfer function) via some filter which alters the reference signal characteristics. For traditional adaptive filtering, the filter of choice is a linear one, whose output $y[i]$ is given by

$$y[i] = \sum_j a_j \tilde{n}[i-j]. \qquad (2)$$

The filter is adapted by adjusting the filter weights a_j so that the filter output $y[i]$ models the primary corrupting noise $n[i]$ as closely as possible. The output of the entire adaptive filter,

$$z[i] = x[i] - y[i] = s[i] + n[i] - \sum_j a_j \tilde{n}[i-j], \qquad (3)$$

will therefore be a close representation of the desired signal $s[i]$. In order to adapt the filter weights so that $y[i]$ models $n[i]$, it is necessary to have two things: (i) a criterion for judging when the model output $y[i]$ is indeed closely modelling the noise component $n[i]$; and (ii) a scheme for using this criterion to adapt the weights of the model to improve the fit.

For the first point, one can easily show (by computing the expectation of the square of the expression for $z[i]$ above) that minimizing the total power of $z[i]$ forces the linear filter output to optimally model $n[i]$.[2] For a perfect model $y[i] = n[i]$, and the total power is minimized and equal to the power in $s[i]$, hence minimizing the total power output also maximizes the SNR of $s[i]$. This general criterion is at the heart of most adaptive filter designs.

An important characteristic of this method is that the minimal power criteria for adapting the filter fit corresponds to a quadratic error function for the error in the signal fit. This is important since the error surface will therefore be quadratic and concave upwards, with a single global minimum, guaranteeing that a minimal solution exists and can be reached by a gradient descent method (although it implies nothing about convergence rates).

The second point above involves utilizing the information from the minimum power criterion to adapt the filter. To accomplish this, nearly all existing methods compute instantaneous gradients of the total power output in terms of the filter weights, and adapt the weights via a simple gradient descent method. Most of the variation in current adaptive methods is in the manner in which the adaptation is accomplished. Perhaps the most common and one of the most powerful methods is the so-called Widrow-Hoff LMS (least mean square) adaptive algorithm. The distinguishing point of this method is that the total squared error (i.e. power) gradient is approximated by the gradient of the instantaneous squared error ($e[i]^2$) between the primary input and the filter output. The rule for updating the filter weights a_j can be written as

$$a_j[i+1] = a_j[i] + \mu \frac{\partial e[i]^2}{\partial a_j} = a_j[i] + 2\mu e[i] \tilde{n}[i-j]. \qquad (4)$$

where $e[i]$ is the instantaneous error $x[i] - y[i]$ and the constant μ controls the speed of convergence. This method has well-defined convergence properties, is numerically very efficient, and has been used in a wide variety of applications. It can been shown that if μ is less than the inverse of λ_{max}, the largest eigenvalue of the autocovariance matrix of the $\tilde{n}(i-j)$'s, convergence is guaranteed. This convergence is typically undesirably slow, however, which leads to methods to improve the choice of μ to speed up convergence.

The principal application of a linear adaptive filter is as a real-time signal component separation (or noise reduction) mechanism. As such, it can be applied to a variety of different problems, with the differences being the way in which the reference signals are defined with respect to the primary signal. Various applications in the literature include separation of tonals, noise, and deterministic signals, as well as process control, voice extraction, spectral line enhancement, notch filtering, and even separation of multiple components by successive tiers of adaptive filters.[5,6,7] Adaptive filters can be constructed in the absence of a separately measured reference signal, where the reference is constructed from a delayed version of the primary signal, chosen so that correlations in the noise and signal components offer effective cancellation. In many cases, adaptive filters provide improvements over static filtering methods when the signals and/or background have time varying or unknown spectral characteristics.

In spite of their wide applicability, adaptive linear filters provide poor performance when the signals in question have significantly nonlinear components, which is the case in a great many real-world situations. Higher order adaptive filters (e.g. quadratic filters) have been shown to perform significantly better than linear filters on signals with nonlinear components. The generalization of this approach to the 'adaptive chaotic filter' can in theory model state-space nonlinearities which would require very high order approximations when modeling solely the scalar time series. This class of filters may thus provide the ability to effectively model reference sources which can appear nearly random to conventional linear filters.

The generalization of global modeling methods to a nonlinear adaptive filter for chaotic and other nonlinear signals is straightforward. In our current approach, we derive a model directly from the input signal (primary input) without the need for a reference signal. This allows for flexibility of the resulting nonlinear adaptive filters. We review the necessary equations and algorithms to demonstrate the above concepts for nonlinear adaptive filters in the next section.

TRADITIONAL ADAPTIVE NONLINEAR FILTERS

Higher-order filter expansions began appearing in the literature in the late 1970s, and were made more practical primarily by the increase in available computational power. Although in principal filters of any order can be used, numerical and computational considerations generally greatly limit the largest order, and most applications of nonlinear adaptive filters involve only a quadratic expansion. The form of a quadratic adaptive filter is a straightforward extension of the linear case -- the only change is the type of terms of the expansion of the filter.[8] The form of a quadratic filter is

$$y[i] = \sum_{j} a_j \bar{n}[i-j] + \sum_{j,k} b_{j,k} \bar{n}[i-j] \bar{n}[i-k], \qquad (5)$$

where the new filter coefficients $b_{j,k}$ correspond to the square and cross terms in the reference signal values and together with the a_j uniquely define the filter. An important aspect of this formulation is that the squared error surface is still a quadratic function, exactly as for the adaptive linear filters. Thus, an equivalent LMS algorithm can be defined to adapt the quadratic filter coefficients via a gradient descent method. The iterative equations for adapting the quadratic coefficients are

$$b_{j,k}[i+1] = b_{j,k}[i] + 2\mu_2 e[i] \bar{n}[i-j] \bar{n}[i-k]. \qquad (6)$$

This equation allows the second order coefficients to be adaptively fit in exactly the same manner as (4), and in fact terms of both orders are fit simultaneously as part of the whole filter. The convergence parameter μ_2 is distinct from the parameter μ for the linear terms, and in general it is necessary to have a separate parameter for each set of higher order terms of the filter. This necessity arises because the adaptive equations for each order contain non-normalized data components, so the instantaneous error gradients for the quadratic terms are proportional to the square of the linear terms, gradients for the cubic terms are proportional to the cube of the linear terms, etc. This has the effect of causing the coefficient corrections for different orders to be of greatly different magnitude, which in turn causes the convergence to be very slow or non-existent.

One way of dealing with the different convergence rates of the various orders of terms in a general filter is to restrict the values of the μ to reflect the difference in possible gradient magnitudes. As an approximate rule, we restrict the value of μ_n for terms of order n so that

$$\mu_n \leq \frac{\mu^n}{n}. \qquad (7)$$

This restriction insures reasonably well that the convergence rates of the various orders do not interfere with the overall convergence of the adaptive filter.

Most of the research on adaptive filters during the last ten years or so has been on techniques for obtaining faster convergence. There are two primary approaches to coefficient estimation: gradient descent methods and least-squares methods. Gradient methods have the advantage of being easy to implement, and are computationally efficient. Least-squares methods are more numerically stable and can have faster convergence rates, however they are often complicated to implement and numerically costly. In addition, there have been recent advances in genetic algorithms and other self-learning algorithms for adaptive filters.[9]

One powerful approach is a simple gradient descent method enhanced via a technique called 'lattice orthogonalization'.[8] This technique rescales the filter contributions not only order by order but also term by term. This is accomplished by computing the autocovariance matrix, defined instantaneously for time i as

$$C_{j,k} = <\tilde{n}[i-j]\tilde{n}[i-k]>, \qquad (8)$$

where the angle brackets denote a time windowed expectation operator. C is then transformed into a diagonal matrix via the Levinson algorithm to obtain the diagonal elements D_m, which are used to re-scale the coordinate terms of the adaptive equations explicitly. For example, the quadratic adaption equation becomes

$$b_{j,k}[i+1] = b_{j,k}[i] + 2\mu_2 e[i] \frac{\tilde{n}[i-j]\tilde{n}[i-k]}{D_j D_k}. \qquad (9)$$

The generalization to other orders is exactly analogous. This scheme has two advantages: first, each term of an nth-order filter is scaled by the exactly the correct average magnitude for the particular combination of coordinate terms; and secondly, because of the explicit scaling only one parameter μ is necessary. The LMS implementation presented later is based on the lattice orthogonalization scheme.

For typical applications, a substantial improvement in convergence rates can be achieved at the expense of numerical stability and computational simplicity by using a matrix update method known as recursive least squares (RLS). While LMS techniques basically find the exact solution for the coefficients at each time step (but only adjust the coefficients a small amount), in RLS formulation, the coefficients are estimates based on a least squares fit over some small time interval. Rather than perform the entire matrix inversion necessary to arrive at the coefficient estimates for each data window, a recursive update formulation for the components of the inverted matrix elements is used. Also, the time window is exponentially weighted, so that past observations are given less importance than more recent observations. The exponential weighting factor is the control parameter which replaces μ in the LMS formulation.[10] We have adapted a relatively stable formulation of the RLS for nonlinear parameter estimation.[11]

The applications discussed in the literature indicate that quadratic filters perform better than linear filters when the signals contain significant nonlinear components.[12,13,14] Although convergence is in general slower than for linear filters, significant increases in output SNR from the filters (as much as 15-20 dB) can be obtained for sufficiently high sampling rates. These results provide a strong argument for the application of higher-order filters to signals which contain strong nonlinearities. Extending this concept, it seems that higher-order filters could yield superior results in general filtering applications, where the applicable signal classes would be much wider. However, the algorithmic and numerical complexity of the direct generalization of the RLS approach makes such filters impractical. Fortunately, results from nonlinear dynamics indicate that it is possible to construct relatively simple models in reconstructed state space that would require very high-order approximations in the time domain. The formulation of filters based on these results is the subject of the next section.

STATE SPACE ADAPTIVE NONLINEAR FILTERS

The equations defining an adaptive state space filter can be developed in several contexts, either assuming the existence of a reference signal in the architecture or not, and using as the underlying model either iterated maps or differential equations. Here we describe a scheme which is a direct modification of our current modelling techniques and which does not assume the existence of a reference signal. The method also allows for considerable flexibility in model design. Inclusion of a reference signal in the algorithm would be a simple modification and may provide a significant improvement in resulting performance.

The following is a brief summary of our existing global nonlinear modeling technique, based on the method of time-delay reconstruction of data. This technique, well known in the literature,[15-18] uses a single scalar variable to reconstruct a 'state space' representation which is topologically equivalent to the phase space of the original system, and for which dynamically invariant measures of the system remain unchanged. The actual method for this reconstruction is straightforward: from a time series $x[t]$ of scalar measurements of some variable, we choose the number of variables d that we require in our state space (i.e. the embedding dimension or number of degrees of freedom), an embedding delay τ, and then we form new d-vectors $y[t]$ by writing

$$y[t] = (x[t], x[t-\tau], x[t-2\tau], \ldots x[t-(d-1)\tau]). \tag{10}$$

The orthogonality of the axes of the new space is approximated by requiring that the delay τ be a minima of a nonlinear correlation measure of the time series, called the mutual information.[19,20] In the absence of an unambiguous result from a mutual information calculation, the time of the first zero-crossing of the autocorrelation function can be used as the delay. While there are alternative methods available for reconstructing a state space, such as genetic algorithm approaches,[21] the foregoing is the most commonly utilized. By reconstructing the state space of a time series in this manner, modeling, prediction and noise reduction can be performed on the (hopefully) simple structure of the state space trajectories rather than the complex behavior of the scalar time series.

One approach to state space modelling is to fit empirical dynamical equations to recover the time evolution of the system.[22-26] The power of this approach is that it allows for a simple, few parameter analytic description which results in great noise averaging characteristics. We have developed an approach, based on fitting empirical maps and ODE's to data, which is numerically efficient. The method allows us to significantly reduce numerical errors in the extraction of coefficients, which decreases the error in prediction and signal processing applications. The method consists of postulating a model for the dynamics of our reconstructed vector data $y[t]$ by writing

$$y[t+1] = \sum_j a_j p_j(y[t]) \text{ or } \frac{dy[t]}{dt} = \sum_j a_j p_j(y[t]) \tag{11}$$

for iterated maps or differential flows, respectively. In (11), the a_j are coefficients and $p_j(y[t])$ is the jth term of some basis (typically a Taylor, Fourier, or Legendre expansion) of the reconstructed state space vector $y[t]$ from (10). In this representation, the nonlinearities of the model are absorbed in the basis set, so we can solve the problem for the a_i as a linear matrix equation.

To develop the equations necessary to define a nonlinear adaptive filter based on an iterative mapping, recall (2), whose output $y[i]$ is then subtracted from the primary signal (3). For a nonlinear map-based filter, we construct an equation based on the state space reconstruction, choosing a dimensionality and order either sufficiently general to model arbitrary signals, or tuned for a particular class of expected signals. In analogy with the state space models discussed above, we write model equation

$$\hat{x}[t+1] = \sum_j a_j p_j(x[t], x[t-\tau], ... x[t-(d-1)\tau]). \tag{12}$$

Note that this representation is similar in form to (2) and (5). The key differences are that we model $x[t]$ directly, rather than model some reference $\bar{n}[t]$, and that we provide for the choice of a tuned dimension d and delay τ to minimize the number of terms required in the filter. To define an LMS adaptive filter, we take the error term for the difference signal

$$e[t] = x[t] - \sum_j a_j p_j(x[t-1], x[t-1-\tau], ... x[t-1-(d-1)\tau]), \tag{13}$$

(note the change from $t+1$ to t on the left hand side) and we write our adaption equation based on the derivative of the square of the error term with respect to the coefficients a_j as for conventional LMS filters. Again, because the error term and

168   Adaptive Nonlinear Dynamical Processing

adaptive equation are analogous to the conventional filter, we expect the error surface to be quadratic and have a single global minima, so that the convergence is smooth and well behaved.

The significant difference between the map-based adaptive filter and a conventional nonlinear filter is the re-interpretation of the coordinate terms of the filter as lying in a multi-dimensional state space of the signal, rather than being simply delayed terms in the time domain. Optimal modelling of a particular signal class can be obtained by picking the time lag of the state space filter according to the auto-correlation time or mutual information time. For an arbitrary signal, this delay time could be calculated continuously and τ adjusted accordingly. To filter a particular signal type, τ would be set to the particular correlation time of that signal, offering a selective modelling criteria. Also, the introduction of general basis functions p_j allows a richer variety of signal behaviors to be modeled using relatively few parameters.

The second formulation of an adaptive state space filter is to use differential flows for the state space model. This formulation is a new and fundamentally different formulation of an adaptive filtering process, using a model comprised of differential equations rather than discrete mappings or time domain values. To derive this type of filter, we follow (11), and write

$$\frac{d\hat{x}[t]}{dt} = \sum_j a_j p_j(x[t], x[t-\tau], \ldots x[t-(d-1)\tau]). \tag{14}$$

As in (13), we let the error function for the LMS algorithm be

$$e[t] = \frac{d\hat{x}[t]}{dt} - \sum_j a_j p_j(x[t], x[t-\tau], \ldots x[t-(d-1)\tau]). \tag{15}$$

The derivatives in (14) and (15) are approximated from the data by taking the slope from a three point exact or five point least squares quadratic fit to each point:

$$\frac{d\hat{x}_3[t]}{dt} = \frac{x[t+\delta t] - x[t-\delta t]}{2\delta t} \quad \text{or} \tag{16a}$$

$$\frac{d\hat{x}_5[t]}{dt} = \frac{2x[t+2\delta t] + x[t+\delta t] - x[t-\delta t] - 2x[t-2\delta t]}{6\delta t}, \tag{16b}$$

depending on the noise level (δ is the sampling interval). Interestingly, the adaptive equations retain the same form as for maps, even though the model is now differential equations. This is because the differential equation formulation is of the same form as a map, except that we are now mapping the local slopes.[24] Previous applications of a stationary (global) version of this method have shown these differential models to have predictive abilities with a number of applications.[27,28]

We now present some numerical results using the methods just described, and will then turn to potential applications in signal detection and classification. These numerical tests were designed to demonstrate two points: (i) that the equations for adaptive modelling of a chaotic signal are valid and that we can indeed use them to extract a state space model adaptively; and (ii) that this type of formulation can track a time varying (non-stationary) signal by continuously adapting the model.

To demonstrate recovery of the proper system coefficients, we used three known chaotic systems. Two of them were iterated maps: the Henon map

$$x[t+1] = 1 - 1.4x[t]^2 + 0.3x[t-1] \tag{17}$$

and the chaotic logistic map

$$x[t+1] = \alpha x[t](1 - x[t]) \tag{18}$$

with $\alpha = 3.7$. We also used the Lorenz system of differential equations:

$$\frac{dx}{dt} = \sigma(y-x) \tag{19a}$$

$$\frac{dy}{dt} = x(R-z) - y \tag{19b}$$

$$\frac{dz}{dt} = xy - bz \tag{19c}$$

with $(\sigma, R, b) = (16, 45.92, 4)$.

We generated data from each system above and added simulated zero-mean Gaussian noise to generate SNRs of 60, 40, and 20 dB (0.1, 1.0, and 10% noise, respectively). For each of the systems, we specified the basis p_j of the nonlinear adaptive filter as the appropriate polynomial expansion (two-dimensional for the Henon map, one-dimensional for the logistic map, three-dimensional for the differential flows, and 2nd order for all cases). Initial coefficient values of zero were used to start the adaptation in every case, and each system was allowed to adapt for 5000 characteristic orbits (orbits are just iterations for Henon and logistics, while they are about 0.5 time units for the Lorenz system). The resulting coefficient vectors were compared to the known ideal values, and the difference between the estimated and the ideal coefficient vectors was expressed as a normalized distance (i.e. a distance of 0.01 means that there was a 1% error in the coefficients). Table 1 summarizes the results of these tests, in which the LMS μ was 0.1, and the high order terms were scaled according to (9). For the RLS algorithm, the weighting parameter was set to 0.95.

Table 1. Adaptive Coefficient Determination Results

System	Normalized Coefficient Vector Error	
	LMS	RLS
Logistic:		
no noise	4.71×10^{-5}	4.71×10^{-5}
60 dB	4.99×10^{-4}	3.50×10^{-4}
40 dB	6.62×10^{-3}	3.97×10^{-3}
20 dB	5.94×10^{-2}	4.30×10^{-2}
Henon:		
no noise	6.16×10^{-5}	6.16×10^{-5}
60 dB	6.18×10^{-4}	4.00×10^{-4}
40 dB	1.20×10^{-2}	3.60×10^{-3}
20 dB	1.01×10^{-1}	7.53×10^{-2}
Lorenz (x coordinate):		
no noise	2.08×10^{-3}	1.61×10^{-2}
60 dB	2.13×10^{-3}	1.60×10^{-2}
40 dB	4.66×10^{-3}	2.05×10^{-2}
20 dB	9.48×10^{-2}	1.03×10^{-1}

Table 1 clearly indicates that we can extract dynamical information from chaotic systems using adaptive methods to a precision on the order of the magnitude of the noise. No attempt was made in these tests to tune the convergence parameters for optimal performance, so it is not unlikely to assume that we could get even better performance with tuning (particularly for noisy flows, for which we also have methods to better estimate the slope that will help as well). It is also apparent that the LMS and RLS results are comparable. We have found that the primary advantage of the RLS method is that we have fewer numeric problems (i.e. the solution seems to be more stable).

In the experiments to demonstrate the ability to track changes in the parameters we used the logistic map defined previously, and the Rossler system:

$$\frac{dx}{dt} = -(y+z) \tag{20a}$$

$$\frac{dy}{dt} = x + ay \tag{20b}$$

$$\frac{dz}{dt} = b + z(x-c), \tag{20c}$$

which exhibits chaotic behavior for the parameters $(a, b, c) = (0.15, 0.20, 10.0)$. For the logistic map, we generated a time series in which α in (18) was constant at 3.7 for 10000 iterations and then abruptly changed to 3.99 for 10000 additional iterations. We used a μ of 0.5. For this experiment, the error $e[t]$ (13) was retained, and a 10 point windowed standard deviation of $e[t]$ was computed for plotting.

Figure 1 shows the windowed standard deviation in the neighborhood of the coefficient transition for the logistic system. It is clear that the filter can track abrupt changes in the coefficients of this system and that the slope of the convergence depends on μ. In noise-free situations, the convergence rates for the LMS and RLS methods are similar when tracking abrupt parameter changes.

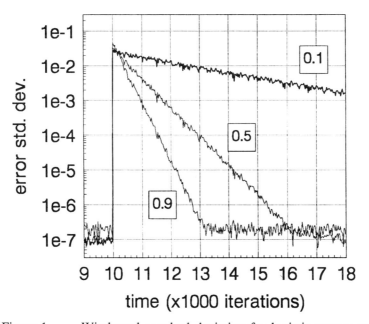

Figure 1. Windowed standard deviation for logistic system near change in α from 3.7 to 3.99 (LMS μ as indicated).

To investigate the ability to track continuously varying parameters, we generated a logistic data set in which α varied sinusoidally between 3.5 and 4.0, with a period of 600 iterations. Figure 2 shows the actual and LMS estimation of α (upper) and the RLS estimation (lower) for the time varying logistic system. A value of 3.75 was used as the initial condition in this case, the LMS parameter was 0.9, and the RLS parameter was 0.6 (these parameters were chosen to give the best performance for this data set). The sinusoidal pattern is clear for the LMS case although a time lag is apparent. The RLS results provide nearly perfect tracking of the time varying coefficient. The superior performance of the RLS method is most likely due to its faster convergence rate.

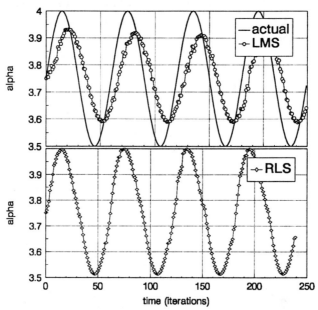

Figure 2. Value of time varying logistic parameter using LMS adaptation (top) and RLS adaptation (bottom).

The RLS adaptive tracking method applies to nonstationary flows as well. We replaced the parameter a in (20b) with a sinusoid with a period of 30 seconds (about 5 orbits of this system) and a range from -0.05 to 0.35. We used this set of ODE's to generate data from a 'nonstationary' Rossler. Using 10000 noise free data points, we obtain the following global (stationary) equations:

$$\frac{d\hat{x}}{dt} = -0.9974y - 0.9931z \qquad (21a)$$

$$\frac{d\hat{y}}{dt} = 0.1273 + 0.9871x + 0.189y + 0.0117z - 0.0042y^2 + 0.0014yz \qquad (21b)$$

$$\frac{d\hat{z}}{dt} = 0.1749 + 0.0441x - 0.9745z + 0.9739xz. \qquad (21c)$$

The normalized RMS of $e[t]$ for these coefficients was 5.49×10^{-3}. The observed value for a (0.189) is near the average value for this nonstationary system. Using the RLS method for parameter tracking, we obtain the estimate of a shown in Figure 3 (the other coefficients, which also vary with time, are not shown here).

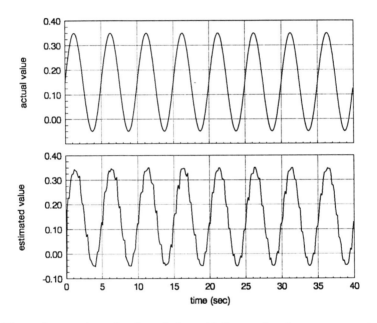

Figure 3. Actual value of sinusoidally varying Lorenz parameter (top), and that estimated using RLS adaptation (bottom).

The normalized RMS of $e[t]$ obtained with the adaptive model is 6.32×10^{-4}, which represents a 19 dB decrease in the residual error. Clearly, we are able to track the nonstationary system parameters for ODE's as well as for maps. Now we turn to a discussion of potential applications of these techniques.

DYNAMICAL MATCHED FILTERS

This section describes two implementations of a 'dynamical matched filter' concept for classification based on adaptive estimation of the underlying signal dynamics. The first method involves analyzing prediction residuals for classification, while the second involves analyses of distributions in coefficient space. The primary assumption underlying both of these concepts are that the SOIs (or the noise, or both) can be modeled in a deterministic fashion. For purely stochastic transient SOIs, traditional statistical methods for detection need to be used, although preliminary NLD-based noise reduction techniques may prove useful. The idea behind our approach is that the state space models can hold information about all orders of nonlinearity in signals, that this information is contained the phase coupling of the nonlinear modes, and that this coupling is exhibited in the topology of the reconstructed state space trajectories. Numerical tests have indicated that changes in this topology with different signal types are quite pronounced. Therefore, there is potential for improved detection and classification based on quantifying the topology of the time-delay reconstructed signal.

An extension of simple energy filters for detection provides a classification mechanism based on matching data segments to a library of dynamical models. The basic form of an energy detector is as follows: given an observed time series $x[t]$ and window halfwidth w, let

$$E_*[t] = \sum_{i=-w}^{i=w} x[t+i]^2 \qquad (22)$$

Values of $E_*[t]$ in excess of some threshold indicate statistical detection of higher energy (or power) events of duration $\approx 2w$.

Now suppose we have obtained a library of models of various SOIs with known forms and coefficients. Given such a library, we can generate new time series based on prediction residuals. Assuming N dynamic models in the library ($F_1, F_2, ..., F_N$), applying each to the observed data for one iteration yields the one-step residual of the nth model:

$$R_n[t] = x[t+1] - F_n(y[t]) \qquad (23)$$

(the form of (23) is slightly different for differential flows, but we will retain the map formulation without loss of generality). Now we form a variance filter on $R_n[t]$ as in (22), and call it $E_n[t]$:

$$E_n[t] = \sum_{i=-w}^{i=w} R_n[t+i]^2 = \sum_{i=-w}^{i=w} \left(x[t+i+1] - F_n(y[t+i])\right)^2 \quad (24)$$

Given two transient types j and k of equivalent magnitude with dynamics described by F_j and F_k, the energy detector outputs (22) will be equivalent. Also, E_j will be less than E_* for signal j, since some portion of the variance is accounted for by the dynamics F_j. E_j will be greater for signal k than for signal j, and may even exceed E_* for signal k since the dynamics F_j may magnify the power in signal k due to the action of F_k. Thus, the difference between $E_n[t]$ and $E_*[t]$,

$$D_n[t] = E_*[t] - E_n[t] \quad (25)$$

provides a measure of the agreement between the data and the library components. A decision scheme based on $D_n[t]$ can be easily constructed: for some appropriate threshold σ_n, the condition $D_n[t] > \sigma_n$ indicates that F_n is very likely operating at time t, while $D_n[t] < -\sigma_n$ indicates that F_n is very likely not operating at time t. Figure 4 shows the outputs of an energy detector (22) and the matched dynamic filters (25) for a short time series with two alternating transient types.

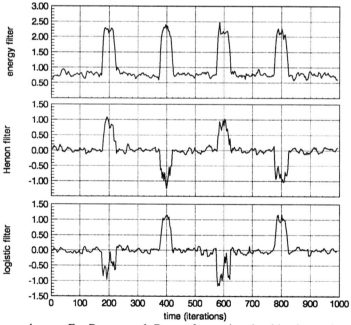

Figure 4. E_*, D_{Henon}, and $D_{logistic}$ for a signal with alternating 75 point Henon and logistic transients (SNR = 8 dB).

Because of the polarity of the output from the dynamic matched filter (we typically observe either no output or a clear positive or negative signal from (25)), the objectives of detection (presence of response) and classification occur simultaneously. We have found that the scheme outlined above works well in cases where the model coefficients are known in advance (or can be computed from the data) and when the input SNR is greater than about 5 dB (55% noise). If the model coefficients are unknown, or if the input SNR is lower than about 5 dB, we use an alternate approach based on partitioning the space of estimated parameters.

The partitioning approach does not require the explicit use of model libraries. Using either a temporally windowed estimate or an adaptive method, an input data set is converted to a vector time series consisting of the estimated coefficients over time. Partitions in this coefficient space are constructed so that coefficient vectors from different signal types are separated as much as possible. Experiments have shown that this approach can yield reasonably high probabilities of detection.

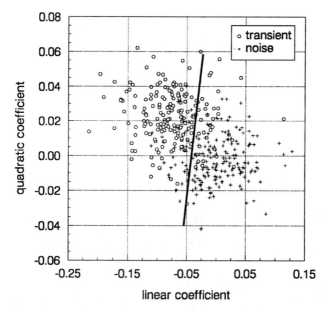

Figure 5. Distribution of two (of five) coefficients for random data for logistic transients. The solid line is the projection of the partition for a pD/pFA of 0.9/0.36 (SNR = -5 dB).

One simple method for constructing partitions is based on computing the geometric center of each distribution of coefficient vectors. Given the centers of two such distributions, create a line passing through them. Any hyperplane perpendicular to this line can serve as a partition. For a particular signal type, experiments can reveal the approximate extent of the region coefficient space it is mapped to under various noise regimes. This information could be used to create multi-variate partitions. As an example, coefficient vectors were derived from data with 400 separate transient bursts (500 points each) of Henon signal at -5 dB SNR (180% noise), and from purely random data with the same statistics. Figure 5 shows the projection of these vectors onto the plane corresponding to the two most significantly different coefficients.

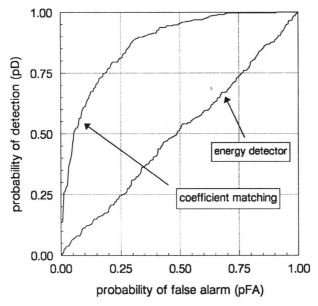

Figure 6. pD/pFA curves from the data in Figure 5 (upper) and from an energy filter approach to transient detection.

Figure 6 shows the probability of detection vs. probability of false alarm (pD/pFA) curves derived from a crude partition of the data in Figure 5 and from a classifier based on an ordinary energy filter of the same data. These curves were generated by sliding the partition through parameter space. The partitions are used as decision rules -- every observation on one side of the partition is declared to be a detection (or classification as a particular type), while those on the other side are ignored (or classified as another type). Since we have estimates of the distributions of the points in parameter space for both actual signals and for noise alone, we can calculate the pD (fraction of actual transient signals correctly identified) and the pFA (fraction of non-signal observations incorrectly classified as signals) as the location of the partition is changed. Notice that at -5 dB, a pD of 0.9 can be achieved at the cost of a pFA of 0.36 using the matched filter approach while the energy detector approach offers almost no advantage.

We conducted additional tests to characterize the performance of the parameter space partitioning method based on noise level, signal duration, and signal type. We used two transient types: Henon and logistic signals. While real world transients are not expected to have such well behaved model descriptions, these are useful examples because for short time windows (on the order of several hundred points or less) the power spectra of these signals are difficult to distinguish. For each signal type, data sets consisting of 500 pairs of alternating signal and silence were created, with varying durations of signal bursts. These 'noise free' signals were then corrupted with simulated Gaussian additive noise at a number of SNR levels. Coefficients were estimated for each of the data segments (signal and non-signal) and pD/PFA curves generated as described above. For every case the pD level at which the pFA first rises above 0.1 was recorded. Figures 7 and 8 show these results for the Henon and logistic transients, respectively.

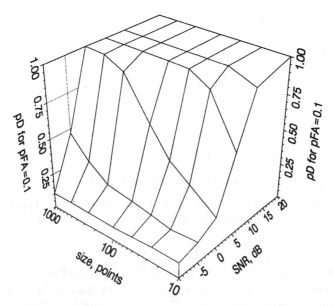

Figure 7. Henon transients: pD achieved for a constant pFA of 0.1 as a function of duration and SNR.

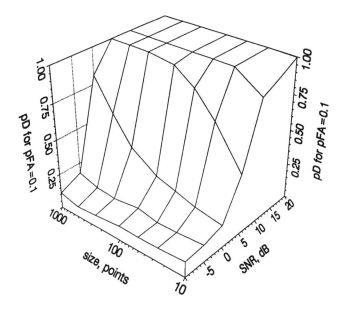

Figure 8. Logistic transients: pD achieved for a constant pFA of 0.1 as a function of duration and SNR.

It is clear that this method can achieve relatively high pDs with a low pFA penalty, even in the presence of considerable amounts of noise. For a given SNR, the longer-lived the transient signal, the better performance we observe. This is to be expected, since more data to fit to a model implies a better estimate of the coefficients. The Henon transients were easier to distinguish from noise (on average the pD values were 0.11 greater at pFA = 0.1) than the logistic. We suspect this is due to the increased size of the coefficient vector (5 vs. 3). These results should be compared with other methods, and the methods presented here need to be applied to real-world signal types.

SUMMARY

We have presented an application of a traditional signal processing method to the estimation of model parameters for stationary and nonstationary nonlinear systems, and demonstrated the ability to extract coefficients in each case. We reiterate that the flow formulation is a fundamentally different approach in adaptive modelling philosophy, in that we fit time varying differential equations to describe the signal as it evolves in the reconstructed state space, with model parameters fit to the derivative of the embedded scalar reference signal. We developed two methods to exploit this parameter estimation capability for signal detection and classification and show that if we can reasonably model a SOI as a nonlinear system, we can detect and classify bursts of these SOIs in very high noise levels. Application areas of these methods include chaotic communication systems; target detection, classification and tracking; fault detection and classification in mechanical systems; biomedical signal processing; and others.

REFERENCES

1. R.J. Roy and J. Sherman, *IEEE Trans. on Auto. Control*, **12**:761-764 (1967).
2. C. Shannon, *Bell Tech. Jour.*, **27**:379-423 (1948).
3. N. Wiener, The Extrapolation, Interpolation and Smoothing of Stationary Time Series with Engineering Applications (Wiley, N.Y., 1949).
4. B. Widrow, J.R. Glover, J.M. McCool, J. Kaunitz, C.S. Williams, R.H. Hearn, J.R. Zeidler, E. Dong and R.C. Goodlin, *Proc. of the IEEE*, **63(12)**:1692-1716 (1975).
5. U. Forssen, *Elect. Lett.*, **26(21)**:1764-1766 (1990).
6. A.C. Orgren, N.R. Malik and D.H. Youn, *Proc. of the ICASSP-86*, **19.4.1**:961-964 (1986).
7. M.R. Sambur, *IEEE Trans. on Acoustics, Speech, and Sig. Proc.*, **26(5)**:419-423 (1978).
8. T. Koh and E.J. Powers, *Proc. of the ICASSP-83*, **1(10)**:37-40 (1983).
9. R. Nambiar, C.K.K. Tang and P. Mars, *Proc. of the ICASSP-92*, **IV**:41-44 (1992).
10. V.J. Mathews, *IEEE Sig. Proc. Mag.*, **8(3)**:10-26 (1991).
11. G.E. Bottomley and S.T. Alexander, *IEEE Trans. on Sig. Proc.*, **39(8)**:1770-1779 (1991).
12. H.M. Sardar and M. Ahmadian, *Trans. of the ASME*, **114**:154-160 (1992).
13. S. Sinha, R. Ramaswamy and J.S. Rao, *Physica D*, **43**:118-128 (1990).
14. D. Zhou, Y. Xi and Z. Zhang, *Int. J. Systems Sci.*, **22(12)**:2563-2571 (1991).
15. D.S. Broomhead and G.P. King, *Physica D*, **20**:217-236 (1986).
16. J.P. Eckmann and D. Ruelle, *Rev. of Modern Phys.*, **57(3)**:617-656 (1985).
17. N.H. Packard, J.P. Crutchfield, J.D. Farmer and R.S. Shaw, *Phys. Rev. Lett.*, **45(9)**:712 (1989).
18. F. Takens, in Dynamical Systems and Turbulence (Springer-Verlag, Berlin, N.Y., 1981), D. Rand and L.S. Young (eds.).
19. A.M. Fraser, Unpublished PhD Thesis, Univ. of Texas at Austin, (1988).
20. A.M. Fraser, *Physica D*, **34**:391-404 (1989).

21. J.L. Breeden and N.H. Packard, N.H, Center for Complex Systems Research Technical Report CCSR-92-11, University of Illinois at Urbana-Champaign (1992).
22. E. Baake, E., M. Baake, H.G. Bock and K.M. Briggs, *Phys. Rev. A*, **45(8)**:5524-5529 (1992).
23. J.L. Breeden and A. Hubler, *Phys. Rev. A*, **42(10)**:5817-4826 (1990).
24. J.S. Brush and J.B. Kadtke, *Proc. of the ICASSP-92*, **V**:321-325 (1992).
25. J. Cremers and A. Huebler, *Z. Naturforschung*, **42(A)**:797 (1987).
26. J.P. Crutchfield and B.S. McNamara, *Complex Systems*, **1**:417-452 (1987).
27. J.B. Kadtke and J.S. Brush, *Proc. of the SPIE Conf. on Sig. Proc., Sensor Fusion, and Target Recog.*, **1699**:338-349 (1992).
28. J.B. Kadtke, J.S. Brush and J. Holzfuss, to appear in *Intl. J. of Bif. and Chaos* (1993).

DETECTION AND DIAGNOSIS OF DYNAMICS IN TIME SERIES DATA: THEORY OF NOISE REDUCTION

Robert Cawley, Guan–Hsong Hsu and Liming W. Salvino
Mathematics and Computations Branch,
Naval Surface Warfare Center, Dahlgren Division, White Oak,
Silver Spring, MD 20906-5640

ABSTRACT

We describe a general four-step approach to chaotic noise reduction: embedding, data state vector alteration, disembedding and iteration. In this way, a noise reduction algorithm may be regarded as a repeated application of an operator $A : v(t) \mapsto \hat{v}(t)$ on a space of scalar time series. We suggest that systematics of the response of a time series to iteration of A can be studied to estimate quantitatively optimal algorithm parameters, such as best embedding trial dimension, $d = d_{pk}$, and number of iterations, $n_M = n_M(d)$, and other quantities, parameters depending on A, to achieve maximum improvement.

INTRODUCTION

When a scalar time series is given, the phase space of an underlying dynamical system can be reconstructed faithfully in a Euclidean space $\mathbb{R}^d$ through certain "embedding" procedures.[1–3] Through this reconstruction, some dynamical or geometrical property can then be used to effect noise reduction. In the application of a chaotic noise reduction algorithm, one necessarily encounters problems of how best to implement it. It is also typical that the procedure can be iterated in some fashion. To achieve best results one needs to choose parameters of the embedding, such as the trial dimension d and time delay Δ for the delay coordinate construction, and the number of iterations to perform, judiciously. For model systems typically studied, such as the Hénon map and Lorenz system with additive noise, the true orbit is known, and one can simply experiment to see what the best choices are. For other types of time series, such as time series with unknown origin, however, such knowledge cannot be obtained through trial and error. In fact, even with model systems contaminated by dynamical noise, typically, the knowledge of a true orbit is also not available. In worse cases, one may not have any clear indication that an underlying dynamics is even there. Thus not only may we not know how best to apply a noise reduction scheme, we likely know nothing about the SNR, S_i, present in the initial data. After noise reduction, even in "friendly", low noise cases, we will still have no idea how well we did, *e.g.*, how much SNR gain we achieved.

In this paper we propose a systematic quantitative approach to the problem of optimal implementation of a chaotic noise reduction algorithm $A : v(t) \mapsto \hat{v}(t)$, on scalar time series data. Our approach is to seek to exploit systematics of the response of $v(t)$ under iteration of A. When an underlying dynamics can be detected, the method should also provide estimates of S_i and of maximal SNR improvement, $\delta_M = \delta_M(d)$. We discuss the results for the case that A represents the local geometric projection (LGP) method first developed by the authors[4,5]. The approach is sufficiently general that it should carry over directly to several other noise reduction methods, with slight modification so they are in the form of maps of spaces of scalar time series to themselves. We explain how to implement this modification below.

In the next section, we describe the four-step process for noise reduction and review the steps briefly for the specific LGP case, $A = A_{LGP}$, in order to fix ideas. We next describe pertinent systematics of performance of A_{LGP} under iteration, and as a function of embedding trial dimension, d. Then we introduce a quantity Σ, which is a function of the original time series, $v(t)$, and that resulting from application of A, viz. $\hat{v}(t)$, alone, and we discuss how to use it. We conclude with a summary at the end.

NEW APPROACH TO CHAOTIC NOISE REDUCTION

A general approach to the noise reduction problem may be seen to proceed through four steps, which exploit the phase space picture of a dynamical system. The LGP method was the first to employ all four steps in a reasonably systematic way. Previous methods have employed variously one, two or three of these four elements. Given a scalar time series, which may be a voltage output of a sensor through which an experimental observation is being made, for example, the four steps are:

(1) represent $v(t)$ through some kind of embedding construction, as a "data-state" vector time series $p(t)$ in a d-dimensional Euclidean space $\mathbb{R}^d$;

(2) through some protocol of "noise reduction," construct from $p(t)$ a "cleaner," candidate data-state vector replacement time series, $\hat{p}(t)$;

(3) disembed, i.e., construct from $\hat{p}(t)$ a suitable scalar replacement time series $\hat{v}(t)$; and

(4) iterate the procedure through maximum noise reduction.

The mathematical idea behind this approach is that, under the dynamical hypothesis on the data, there is a true noise-free orbit contained in an m-dimensional manifold M. The embedding construction, under which $M \to M' \subset \mathbb{R}^d$, should give a complete representation of the orbit dynamics provided d is large enough that an embedding actually results. The observed quantity $v(t)$ is then a noisy version of one of the coordinates for the system.

Most noise reduction methods so far have centered around various approaches to step (2). It has also been recognized that some noise reduction may be achieved right away in step (1) by what amounts to suitable coordinate choices. Except in Refs.[4,5], scant notice has been paid the role of step (3); and likewise step (4) has received little or no attention except as an end in itself. We describe the above four-step process for the LGP noise reduction method next.

Step 1. Embedding construction. We consider first the noise-free chaotic case where the observation $v(t) = V(t)$. To have a complete representation of the chaotic dynamics producing $V(t)$ it is not necessary to measure all of the variables of the system. One possibility may be to choose as coordinates for the phase space, $(V(t), \dot{V}(t), \ddot{V}(t))$, say, where we have assumed three dimensions. This was the state space reconstruction studied originally by Packard, et al.[2]. However, the numerical differentiation of observational data required for this method amplifies high frequency noise that inevitably must contaminate $V(t)$. Another approach, not suffering from this problem, appears first to have been suggested by D. Ruelle (see Ref.[2]), namely, the delay coordinate construction (DCC). Here the coordinate choice for three dimensions is $(V(t), V(t+\Delta), V(t+2\Delta))$, where $\Delta \neq 0$ is the delay. In each of these examples the data scalar $V(t)$ has been converted to a three-dimensional data-state vector.

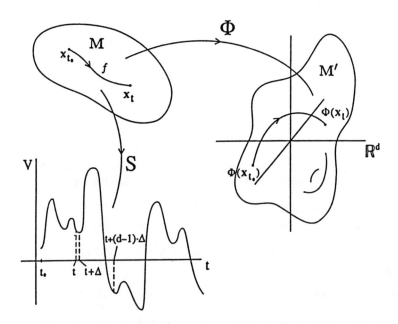

Fig. 1. The Ruelle–Takens delay coordinate construction. from the flow of The scalar function S, in general, is the connection between the system under observation, which mathematically lies in M, and the experimental observation itself, $V(t)$.

Mathematical justification for each of these procedures was provided by Takens.[1] Takens' Theorem says that the map Φ from the m–dimensional manifold M for the dynamics reflected in $V(t)$, into the d–dimensional Euclidean Space $\mathbb{R}^d$ specified by the coordinate choices using $V(t)$ along with its derivatives ($d-1$ of them) or the delays (up to $(d-1)\Delta$), is very generally going to be an embedding (*i.e.*, "perfect") as long as $d > 2m$. The data-state vector orbit in $\mathbb{R}^d$ will thus be contained wholly in the embedded image $M' = \Phi(M)$. Since we may not know m experimentally we do not generally know whether an embedding will result for a given choice of d. However, more recent mathematical work has provided a considerable softening of the restriction on d to $d > 2d_\Lambda$ where d_Λ is the fractal dimension of the attractor Λ holding the data-state vector orbit.[3]

Thus, while in fact an experimental observation is always somewhat noisy, so that

$$v(t) = V(t) + \epsilon \eta(t), \qquad \epsilon > 0, \|\eta\| = 1, \tag{1}$$

where $\|\cdot\|$ denotes the RMS norm and $\eta(t)$ is a normalized random process representing the noise, the DCC can at least provide an embedding for the deterministic part. The main features of the Ruelle–Takens construction are depicted in Figure 1.

Step 2. *"Official"* noise reduction. This is where most work on noise reduction has been focussed. It is also here that various noise reduction algorithms are supposed to differ from one another. But it should be stressed that significant noise reduction can also occur in Step 3 (see below)!

The idea of the LGP method is very simple. For sufficiently large d, all the dynamics implicit in $V(t)$ lies in M', but the noise contribution to the data-state vector for $v(t)$, *i.e.*, to

$$\boldsymbol{p}(t) = (v(t), v(t+\Delta), \ldots, v(t+(d-1)\Delta)), \tag{2}$$

explores every possible direction in $\mathbb{R}^d$ including those lying off of M'. So, we just estimate where M' is from the noisy data, $\boldsymbol{p}(t)$, and move the points of $\boldsymbol{p}(t)$ towards our best guess for M'. Thus $\boldsymbol{p}(t) \to \hat{\boldsymbol{p}}(t)$, the "cleaned-up" vector time series.

The problem of locating M' is one of statistical estimation. Although we may have no idea of the shape of M', whether it is some sort of distorted donut or hyper-torus, a garbly-looking multi-dimensional ellipsoid, or a funny saddle, in all cases if we look only at a sufficiently small cloud of neighboring points, this small part will look flat. That is exactly what it means for M' to be a (differentiable) manifold, namely to be locally flat. So, our strategy is to estimate local hyperplanes tangent to M'. These are the so–called tangent spaces. Since there will also be some small non-flatness (see Figure 2), we don't specify $\hat{\boldsymbol{p}}(t)$ by moving the noisy points of $\boldsymbol{p}(t)$ into these estimated tangent spaces, but only halfway towards them.

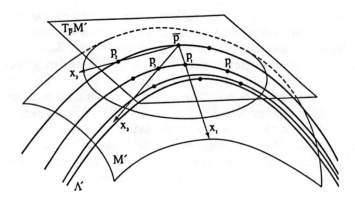

Fig. 2. Tangent plane $T_{\bar{p}}M'$ to M' at a point $\bar{p}$. Λ' refers to the image of the strange attractor Λ on which the orbit under observation is assumed to lie. Points of noisy orbits will typically lie out of M'.

In estimating local tangent spaces of M', an important and consequential fact deserves note. We observe that m may not be known. Since the dimension of the hyper-plane for the projection, k, is a parameter of A_{LGP}, we might have either $k \geq m$ or $k < m$. We have examined the effect of repeated application of the LGP algorithm to clean chaotic time series for both cases[6]. For the Lorenz system, $\dot{x} = 10(y - x)$, $\dot{y} = 28x - y - xz$, $\dot{z} = -\frac{8}{3}z + xy$, with $k = m = 3$, for example, we found good attractor stability; after 256 iterates there was little discernible attractor deterioration. The same was true for $k = 4$. But for $k = 2$, moderate distortion was visible after 64 iterates, and substantial deterioration set in after 256 iterates. We expect in some general way that if $k < m$, or possibly $k < 1 + [d_A]$, where $[d_A]$ is the integer part of the attractor dimension, the attractor will be stable under iteration of A, whereas, for larger k we expect attractor stability.

Since any algorithm must possess some baseline noise, even if $k \geq m$ we should expect eventual instability. Exactly where this happens, i.e., after how many iterates, will be system dependent. So we would like to have some quantitative measure of the effect of attractor stability and instability. These ideas form the basis of the present approach, and we describe one possibility in this paper.

Step 3. Disembedding. Although $p(t)$ is a delay coordinate construction by definition (eq.(2)), the cleaned up version is almost surely not. But the right answer, viz.

$$P(t) = (V(t), V(t+\Delta), V(t+\Delta), \ldots, V(t+(d-1)\Delta)), \qquad (3)$$

certainly is. Thus we seek a best scalar time series $\hat{v}(t)$ that will produce a data-state vector orbit as close as possible to $\hat{p}(t)$. This $\hat{v}(t)$ is our first guess at $V(t)$. To do this we minimize the error,

$$\mathcal{E} = \sum_t \sum_{j=1}^{d} [\hat{p}_j(t) - \hat{v}(t + (j-1)\Delta)]^2 \tag{4}$$

Except for t-values near the end-points of the sum, the minimizing $\hat{v}(t)$ is given by

$$\hat{v}(t) = \frac{1}{d} \sum_{j=1}^{d} \hat{p}_j(t - (j-1)\Delta). \tag{5}$$

Similar expressions result for end-point range t-values[4].

Notice if $\hat{p}(t)$ were actually $P(t)$ as shown in eq.(3), then $\hat{v}(t)$ in eq.(5) would become $V(t)$ as we should require. Since there will surely be some noise still present in $\hat{p}(t)$, we can imagine that the averaging that occurs in eq.(5) will provide further noise reduction. This expectation is borne out in numerical computations. Notice also the disembedding step does not depend on the noise reduction method employed in Step 2, and can just as well be applied, e.g., after a Kostelich–Yorke trajectory adjustment[7].

Summarizing the effects of steps (1) — (3), we have that $v(t) \to p(t) \to \hat{p}(t) \to \hat{v}(t)$, that is, the LGP algorithm A has mapped the scalar time series $v(t)$ to a cleaned-up scalar time series $\hat{v}(t)$, i.e., $A : v(t) \to \hat{v}(t)$. We are ready now for

Step 4. Iteration through maximum improvement. We define the improvement in terms of signal-to-noise ratio (SNR). The initial SNR in db is

$$S_i = 20 \log \frac{\|V(t)\|}{\epsilon \|\eta(t)\|}. \tag{6}$$

Following eq.(1), we write the cleaned up time series as

$$\hat{v}(t) = V(t) + \epsilon \hat{\eta}(t), \tag{7}$$

so the output SNR in db is

$$S = 20 \log \frac{\|V(t)\|}{\epsilon \|\hat{\eta}(t)\|}. \tag{8}$$

Since ϵ can measure the strength of the noise, we simply normalize $\eta(t)$ to suit our convenience. We take $\|\eta(t)\| = 1$. Then the SNR improvement is

$$\delta = S - S_i = 20 \log \frac{1}{\|\hat{\eta}(t)\|}, \tag{9}$$

which depends only on the relative output noise power. Under n iterates, $v(t) \to \hat{v}_n(t)$, with $\hat{v}_n(t) \equiv v_{n+1}(t)$, and $\delta \to \delta_n$.

PERFORMANCE SYSTEMATICS FOR A_{LGP}

As A_{LGP} is iterated, δ typically increases with n until reaches a maximum value, δ_M, after n_M iterates. After reaching the maximum, it falls relatively slowly, but steadily, under further iteration. The values achieved vary with the embedding trial dimension d, the dimension k of the tangent spaces used to estimate M' and the number of points $\nu+1$ chosen to model the local neighborhoods of M'. δ_M also depends on the delay Δ chosen for the DCC, the sampling time ΔT for the case of flow data, the initial noise fraction $\mathcal{N}$, whether it is additive or dynamical noise, the time series length, and the system itself. We have performed extensive systematic investigations of almost all these dependences, much of which is reported in References.[4,5] We show here only a representative sample (Figure 3).

We note a few pertinent features of the plots shown : (a) the variation of δ under iteration is very slow, especially after reaching maximum. Accordingly, we use

$$\alpha = \log_2 n \tag{10}$$

as the iteration variable. For each d, δ is approximately linear initially and

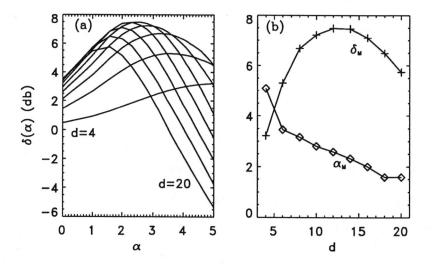

Fig. 3. The example shown is for noisy Lorenz time series with $S_i \doteq 20db$ (10% white noise). The sampling time is 0.05 and the delay $\Delta = 4$. $(k, \nu) = (3, 40)$ and $d = 4, 6, 8, \ldots, 20$. (a) For each trial dimension d, SNR improvement $\delta(\alpha)$ typically increases with initial application of the LGP algorithm then falls after reaching a maximum $\delta_M(d)$ at $\alpha_M(d)$ for each d. (b) +—$\delta_M(d)$, ◇—$\alpha_M(d)$.

roughly parabolic near maximum at $\alpha = \alpha_M = \log_2 n_M$. (b) δ_M rises with d, peaking at $d = d_{pk}$ and falls subsequently with further increase of d; and (c) α_M typically falls roughly linearly as d rises.

We remarked earlier that steps (1) and (2) have received the most attention in noise reduction efforts, but with little attention paid to steps (3) and (4). We addressed the role of step (3) in the previous section. We next investigate how one might exploit the systematics of Step 4 just described.

$\Sigma = \Sigma(\alpha)$, A DATA OBSERVABLE

We assume that the true signal V and initial noise η are uncorrelated, *i.e.*,

$$C_{V\eta} = \|V\|^{-1} <V, \eta> = 0, \tag{11}$$

where $<\cdot,\cdot>$ denotes the inner product, *e.g.*, $<V,\eta> = \frac{1}{N_D} \sum_{t=1}^{N_D} V(t)\eta(t)$. We consider the quantity

$$\Sigma = 20 \log_{10} \frac{\|v(t)\|}{\|v(t) - \hat{v}(t)\|}. \tag{12}$$

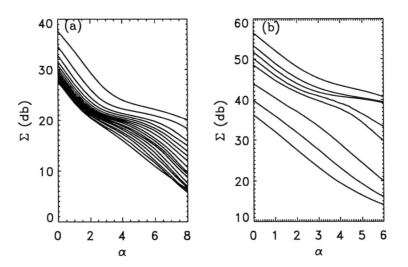

Fig. 4. $\Sigma(\alpha)$ for (a) the Lorenz system, for the same set of runs shown in Figure 3; (b) the Hénon map with 1% initial noise ($S_i \doteq 40\text{db}$). Orbit length $N = N_D - (d-1)\Delta = 3000$ points, $(k, \nu) = (3, 40)$, $\Delta = 1$. $\Sigma(\alpha)$ shown is an average of ten runs (full iteration sequences up to 128) of the algorithm applied to the same noisy orbit time series. The plateau occurs at $\Sigma \sim S_i$, the initial SNR, and can be seen clearly for both flows and maps. This feature is absent for purely random time series.

From eqs.(1) and (7),

$$\Sigma = \Sigma_i - 20\log_{10}(1 + \|\hat{\eta}\|^2 - 2\|\hat{\eta}\|C)^{\frac{1}{2}} \qquad (13)$$

where

$$\Sigma_i = 20\log_{10}\left(\frac{\|V + \epsilon\eta\|^2}{\epsilon^2}\right)^{\frac{1}{2}} = 20\log_{10}\left(\frac{\|V\|^2}{\epsilon^2} + 1\right)^{\frac{1}{2}} \qquad (14)$$

and

$$C \equiv C_{\eta\hat{\eta}} = \|\hat{\eta}\|^{-1} <\eta, \hat{\eta}>. \qquad (15)$$

From eq.(14), we note that for ϵ small, $\Sigma_i \sim S_i = 20\log_{10}\frac{\|V\|}{\epsilon}$.

Initially, $\hat{\eta}_0(t) \equiv \eta_1(t)$, hence $\eta - \hat{\eta}$ vanishes and Σ is infinite. As α rises towards α_M, Σ falls, and approaches Σ_i and hence S_i. This assumes $\hat{\eta}(t)$ can be neglected in the second term of eq.(13) when $\alpha \simeq \alpha_M$. In an idealized case when $\hat{\eta}$ is perfectly correlated with η after $\alpha = \alpha_M$, $C = 1$ and $\Sigma = \Sigma_i - 20\log_{10}(1-\|\hat{\eta}\|)$. As α rises above α_M, Σ begins to rise again so that its minimum would then provide an estimate of S_i and the location of α_M. However, $C = 1$ only at iteration zero ($\alpha = -\infty$) and, in fact, in all our numerical studies, decreases monotonically toward zero as α increases. Moreover, since $\|\hat{\eta}\|$ is increasing for $\alpha > \alpha_M$, the competition between these two terms makes it unclear what behavior Σ should have if eq.(13) is all we have to go on. What does happen, in every example we have studied, is illustrated in Figure 4 for the Lorenz system and the Hénon map, $x' = 1.4 - x^2 + 0.3y$, $y' = x$.

Apart from the additive constant Σ_i, Σ is a function of $\|\hat{\eta}\| = \|\hat{\eta}\|(\alpha)$ and $C = C_{\eta\hat{\eta}}(\alpha)$. Up to an obvious factor of $20\log_{10} e$ we may write

$$\Sigma(\alpha) = \Sigma_i + \frac{1}{2}\log\frac{1}{D(\alpha)}, \qquad (16)$$

and formally, since α is defined only at discrete values, $\alpha = \log_2 n$, $n = 1, 2, \ldots$,

$$\Sigma'(\alpha) = d\Sigma(\alpha)/d\alpha = N(\alpha)/D(\alpha), \qquad (17)$$

with

$$N(\alpha) = \|\hat{\eta}\|(\alpha)\{[\|\hat{\eta}\|(\alpha) - C(\alpha)]\delta'(\alpha) + C'(\alpha)\} \qquad (18)$$
$$D(\alpha) = [\|\hat{\eta}\|(\alpha) - C(\alpha)]^2 + [1 - (C(\alpha))^2]^2. \qquad (19)$$

Eqs.(9) and (16)—(19) suggest that if forms of $\|\hat{\eta}\|$ and C are known for a given algorithm, then the analysis of $\Sigma(\alpha)$ and $\Sigma'(\alpha)$ should provide quantitative information about the given time series, and about algorithm performance. Thus note that for a given d, $\delta(\alpha)$ may be regarded as a function of $\delta_M = \delta_M(d)$ and $\alpha_M = \alpha_M(d)$. We expect this information to be present in the behavior of $\Sigma(\alpha)$.

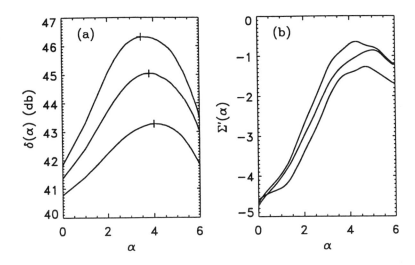

Fig. 5. (a) *SNR* improvement for the noisy Hénon time series example of Figure 4(b) for trial embedding dimensions from 4, 5 and 6. Values shown here for each α, like those of Figure 4(b), are averages over the same ten runs of A_{LGP}. (b) $\Sigma'(\alpha) = d\Sigma/d\alpha$ from the data of Figure 4(b). The derivative was computed from $\Sigma(\alpha)$ from values defined for non–integer n by interpolation. For random time series, $\Sigma(\alpha)$ does not display the characteristic peak structure evidenced here.

Indeed the peaking behavior of $\Sigma'(\alpha)$ does seem to correspond qualitatively to that of $\delta(\alpha)$ (*cf.* Figure 5).

In fact, using a very simple phenomenological model for $\delta(\alpha)$ and $C(\alpha)$ we have been able to obtain the quantities δ_M and α_M reasonably successfully for several examples. We show some preliminary results in Figure 6.

SUMMARY

We have proposed a new iterated approach to chaotic noise reduction that can be described as a four–step procedure. We have briefly described each of the four steps and discussed the role of each in the overall method. In particular, the method shows promise as a quantitative basis for optimal choice of algorithm parameters. We have illustrated the method for the LGP algorithm, but observed that it should be applicable to most other phase space based methods.

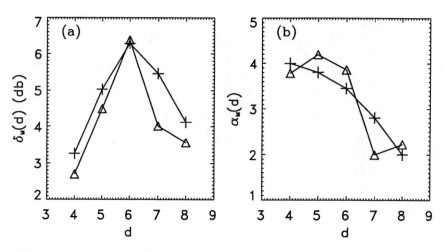

Fig. 6. $\delta_M(d)$ and $\alpha_M(d)$ for the data of Figure 4(b): $+$ — results of applications of A_{LGP}. $\triangle$ — estimates from $\Sigma'(\alpha)$ via model.

ACKNOWLEDGMENTS

This work was supported by the Office of Naval Research and the Independent Research Program of Naval Surface Warfare Center, Dahlgren Division.

REFERENECES

1. F. Takens, in *Dynamical System and Turbulence, Warick 1980*, No. 898 in *Lecture Notes in Mathematics*, edited by D. A. Rand and L.-S. Young (Springer–Verlag, Berlin, New York, Berlin, 1981), pp. 366–381.

2. N. Packard, J. Crutchfield, J. Farmer, and R. Shaw, Phys. Rev. Lett. **45**, 712 (1980).

3. T. Sauer, J. A. Yorke, and M. Casdagli, J. Stat. Phys. **65**, 579 (1991).

4. R. Cawley and G.-H. Hsu, Physical Review A **46**, 3057 (1992).

5. R. Cawley and G.-H. Hsu, Phys. Lett. A **166**, 188 (1992).

6. R. Cawley and G.-H. Hsu, in *Proceedings of the 1st Experimental Chaos Conference*, edited by S. Vohra et al. (World Scientific, Singapore, 1992), pp. 38–46.

7. E. Kostelich and J. A. Yorke, Phys. Rev. A **38**, 1649 (1988).

CHAOTIC NOISE REDUCTION BY LOCAL–GEOMETRIC–PROJECTION WITH A REFERENCE TIME SERIES

Robert Cawley and Guan–Hsong Hsu
Mathematics and Computations Branch,
Naval Surface Warfare Center, Dahlgren Division, White Oak,
Silver Spring, MD 20903-5640

ABSTRACT

We describe a family of algorithms for chaotic noise reduction when a "reference" time series is assumed to be given. These algorithms are variations of the local–geometric–projection (LGP) algorithm first introduced in[1,2]. Signal-to-noise ratio (SNR) improvements achieved depend on the quality of the reference time series, the degree to which information available from knowledge of the reference data is used, and on the SNR initially present in the data. We report results for cases where the reference time series is the original noise-free time series, and where it is taken to be a noise reduced version of the given noisy time series.

INTRODUCTION

In previous work[1,2], the authors introduced a local geometric projection (LGP) algorithm to perform noise reduction for chaotic time series. The method is based solely on the geometric properties of dynamical systems. It was demonstrated that the LGP method is effective for both maps and flows, including coarsely sampled cases, and achieves typically 10–$15 dB$ signal–to–noise ratio (SNR) gain for $0 dB$ initial SNR in the case of flows.

The LGP method does not make use of specific prior dynamical information, and, indeed, can equally well be applied to pure noise time series. On the other hand, some noise reduction methods for chaotic data do make use of quite concrete *a priori* knowledge. For example, the shadowing methods[3,4] assume complete knowledge of a map, while a probabilistic method[5] assumes a noise–free orbit realization to be given. Although this kind of information seems unlikely to be available for data of unknown origin, in general, it might be in some cases that it is. The *SNR* gain when this is the case can be significantly better than the fairly typical 10–15dB seen by us, and by other researchers using other noise reduction methods[6–10].

Possibly more importantly, there is also a theoretical reason for investigating a chaotic noise reduction algorithm when sample reference data are available. Used in conjunction with a given algorithm's baseline noise properties, *i.e.* the noise introduced by its application to originally noise free data, performance of the algorithm when reference time series information is exploited can provide insight as well as data regarding theoretical limits as to how well a given method is capable of performing ideally. In this note, we content ourselves with investigating a few variations of the LGP approach when we assume knowledge of a reference time series to be available.

In the next section, we describe two primary classes of reference data in a dynamical system and review briefly two variants of the LGP method. Next we describe three "levels" of reference time series use. We present results of comparison of Level 1 against the nominal LGP algorithm for various amounts of initial noise for both variants. We also present selected results for two examples of reference time series, first where the reference time series is the original noise–free time series, and second where a noise–reduced version of the given noisy time series is used. Finally, we offer preliminary results for Levels 2 and 3 use of reference time series. We conclude the paper with a summary and discussion.

DESCRIPTION OF THE LGP METHOD

Let $v(t)$ denote a time series contaminated by noise,

$$v(t) = V(t) + \epsilon \eta(t), \quad t = 1, 2, \ldots, N_D, \tag{1}$$

where $V(t)$ is the true, noise–free time series, and $\eta(t)$ is the noise process, with $\|\eta\| = 1$, so that ϵ is proportional to the noise fraction. We assume that there is a smooth dynamical system,

$$f^s : M \to M, s \in \mathbb{R} \text{ or } \mathbb{Z}, \tag{2}$$

where M is the phase space. The true time series $V(t)$ is a sequence of discrete measurements of some scalar quantity $S : M \to \mathbb{R}$ along an orbit, *viz.*,

$$V(t) = S(f^{s_0 + (t-1)\Delta s}(y_0)), \tag{3}$$

for some initial point $y_0 \in M$, starting from some initial time s_0 (typically taken to be 0), and sampling time Δs. For maps one normally takes $\Delta s = 1$.

There are two main classes of reference time series for $v(t)$. In the first, the reference time series coincides with the given noise free data $V(t)$, or possibly a noisy version of this. The second class consists of an infinite collection of time series generated from different orbit realizations from the same dynamical system on the same attractor, or a noisy version of this. For example, one might have a collection of sample orbits from a system one is interested in. In this paper, we consider only the first class of reference orbit. Methods similar to those we use here are easily devised for the second class.

In the first class, where we have a noisy version of $V(t)$, we might have come by it a number of ways. For finely sampled flow data, for instance, it might have been produced by a suitable low passing of the given time series such as in Sauer's implementation of the LGP idea[8]. Alternatively, it might have been produced by the chaotic noise reduction algorithm itself, previously applied without use of any reference data.

The LGP algorithm is an iterative procedure. Each iterate consists of three steps :

(1) Embedding-- represent $v(t)$ through some kind of embedding construction, as a "data-state" vector time series $p(t)$ in a d-dimensional Euclidean space $\mathbb{R}^d$;

(2) Data–state vector adjustment-- through some protocol of "noise reduction," construct from $p(t)$ a "cleaner," candidate data-state vector replacement time series, $\hat{p}(t)$;

(3) Disembedding-- from $\hat{p}(t)$ construct a "best" scalar time series $\hat{v}(t)$ to replace $v(t)$.

We use reference orbit information in step (2). Step (1) is fairly standard for phase space methods. One can use either the delay coordinate construction[11], or its generalization due to Sauer et al.[12]. Step (3) was first discussed by the authors in Refs.[1,2].

We elaborate the "official" noise reduction step (2): For sufficiently large d, all the dynamics implicit in $V(t)$ lies in M', but the noise causes the vector

$$p(t) = (v(t), v(t+\Delta), \ldots, v(t+(d-1)\Delta)) \qquad (4)$$

to explore all directions in $\mathbb{R}^d$, including those lying off of M'. So, we just estimate where M' is from the noisy data, $p(t)$, and move the points of $p(t)$ toward our best guess for M'. Thus $p(t) \to \hat{p}(t)$, the "cleaned-up" vector time series.

The problem of locating M' is one of statistical estimation. Although we may have no idea of the shape of M', locally everywhere it is well approximated by its tangent space. So, our strategy is to estimate, for each of a suitable collection of points, the local hyper-plane tangent to M', or a (normally) larger linear subset of $\mathbb{R}^d$ based at the same point. First we estimate base points, as follows.

We choose a point $p_0 \in \mathbb{R}^d$ from the noisy orbit and its ν nearest neighbors, $p_1, \ldots, p_\nu$. The centroid of this ball of points is $q = \frac{1}{\nu+1} \sum_{i=0}^{\nu} p_i$, which is our estimate. From the points of the neighborhood corresponding to p_0, we form the unit vectors,

$$x_i = \frac{p_i - q}{\|p_i - q\|}, \quad i = 0, 1, \ldots, \nu, \tag{5}$$

from which we next estimate a best k-dimensional linear space H approximating $\{x_i\}$, where k is chosen arbitrarily[1]. The affine linear space $q + H$ is then an estimate of the best k-dimensional linear space containing the tangent space for the chosen point, p_0. The noise reduction for the points of the time series, $p_i, i = 0, \ldots, \nu$, is then achieved by projecting them towards the space $q + H$. A cover of the phase portrait is constructed by randomly selecting the center points p_0, and the projection procedure is repeated for each neighborhood. For details see Reference[1].

In order to take into account effects associated with curvature of M' in a crude way in low noise cases, we have used a fraction parameter, $0 \le f < 1$, to limit distances we move the noisy points towards the estimated spaces, $q + H$. This was intended also to soften effects of statistical outliers in high noise cases. $f = 0.5$ means half-way to $q + H$, and $f = 0.0$ means simple projection into $q + H$.

THREE ALGORITHMS USING REFERENCE TIME SERIES

We suppose we have a reference time series $\tilde{v}(t)$ which is "better" in some sense than $v(t)$. Using the same embedding procedure we get a new vector time

---

[1] k is a parameter of the algorithm since, in general, we may not know the topological dimension, $m = dimM = dimM'$. Only for $k = m$ is it that $q + H$ is actually a tangent space estimate for M'.

series $\tilde{p}(t)$ which is the reference orbit. We define four levels of use of the reference data, which here are based on the assumption that $V(t)$ and $\tilde{v}(t)$ are either identical or close:

Level 0: Do not use the reference data. Apply LGP itself, unmodified.

Level 1: Given $p_0 = p_0(t_0)$, find the corresponding point $\tilde{p}_0$ in time in the reference orbit and find the ν nearest neighbors for $\tilde{p}_0$ on $\tilde{p}(t)$, viz. $\tilde{p}_i = \tilde{p}(t_i)$ for $i = 1,\ldots,\nu$. This is the only use of the reference time series. Then take $p_i = p(t_i)$, $i = 0,\ldots,\nu$, and apply the remaining LGP steps to these points.

Remark: Level 1 only uses the reference orbit to locate nearest neighbors. In high noise cases, e.g., 0dB, the nearest neighbors found for a chosen point p_0 by searching on $\{p(t)\}$ might be significantly different from those found if there were no noise. Thus, one might expect a cleaner reference orbit to help find nearest neighbors more correctly. Taking $\tilde{v}(t) \equiv V(t)$ gives an idea of how much improvement can come from just locating nearest neighbors correctly.

Level 2: Proceed as in Level 1 and estimate the linear space H exactly as described above. But now, instead of taking $q + H$ as our local linear space estimate corresponding to p_0, we take $\tilde{q} + H$, a translation of H to a new, "improved" centroid $\tilde{q} = \frac{1}{\nu+1}\sum_{i=0}^{\nu}\tilde{p}_i$, determined by the reference orbit.

Level 3: Use the reference orbit to estimate the whole linear space : find $\tilde{p}_i$, $i = 0, 1, \ldots, \nu$, and $\tilde{q}$ as in Levels 1 and 2, respectively; estimate the linear space $\tilde{H}$ from unit vectors

$$x_i = \frac{\tilde{p}_i - \tilde{q}}{\|\tilde{p}_i - \tilde{q}\|}, \quad i = 0, 1, \ldots, \nu. \tag{6}$$

Then project the p_i towards $\tilde{q} + \tilde{H}$ to effect the noise reduction. We remark that this estimate coincides with that issuing from initial (*i.e.* first iterate) application of the LGP algorithm to noise–free time series. We also note that k is still a free parameter.

PERFORMANCE

For our studies we use the Lorenz system,

$$\begin{aligned} \dot{x} &= \sigma(y - x), \\ \dot{y} &= \rho x - y - xz, \quad (x,y,z) \in \mathbb{R}^3 \\ \dot{z} &= -\beta z + xy, \end{aligned} \tag{7}$$

$$\tag{8}$$

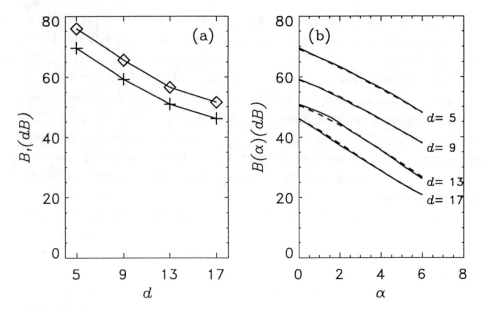

FIG. 1. Lorenz data: (a) First iterate baseline noise comparison for Level 0, $f = 0.0$ (+) and $f = 0.5$ (◊). (b) Baseline noise for Level 0 (solid line) and Level 1 (dashed line) with $f = 0.0$.

with additive white noise. Our time series is drawn from a solution $x(t)$ obtained by a fourth order Runge–Kutta integration of Eqs.(7) with integration step size 0.005 and sampling time $\Delta t = 0.05$ (17.5 points/typical oscillation). For algorithm parameters we choose $(k, \nu) = (3, 40)$, and, except for a few (noted) instances, we take $\Delta = 2$.

We begin with baseline noise calculation results. This is the noise introduced into originally noise–free data through application of the algorithm. In addition to k, ν, and Δ, the baseline noise depends on other algorithm parameters, such as $f, d,$ and n, the number of iterates. We represent the effect as a signal-to-noise ratio,

$$B_n = 20 \log \frac{\|V\|}{\|\hat{v}_n - V\|} \tag{9}$$

where $\hat{v}_n = \hat{v}_n(t)$ is the output time series from n iterates.

Numerically, we find that the single iterate ($n = 1$) baseline noise falls monotonically with increasing d, exhibiting substantial variation as it does so (Figure 1(a)). For $d = 15$, $B_1(f = 0.0) = 48.5 dB$ from the figure, while $B_1(f = 0.5) = 53 dB$. So for the present noise free Lorenz case the $f = 0.0$ version

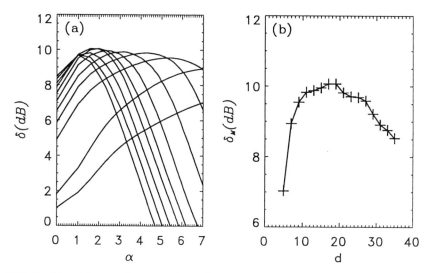

FIG. 2. Algorithm performance for $20dB$ noisy Lorenz data for Level 0 and $f = 0.0$: (a) SNR gain $\delta(\alpha)$ for $d = 5, 7, \ldots, 25$; (b) maximum gain $\delta_M(d)$.

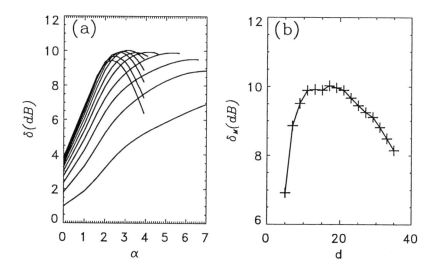

FIG. 3. Same as Figure 2 but $f = 0.5$.

of the algorithm is about $4.5dB$ noisier than the $f = 0.5$ version. For further comparison we have $B_1(f = 0.5) = 68dB$ for delay $\Delta = 1$, which is $15dB$ "quieter". Regardless of algorithm parameter values, however, we note that B_1 must be the same for each of Levels 0–3. In fact, we find something much stronger for Levels 0 and 1, viz., no significant difference for all n (Figure 1(b)). The small differences shown at each $\alpha = \log_2 n$ are not significant since algorithm performance exhibits about a tenth of a dB fluctuation owing to the randomness of the local neighborhood selection procedure in covering the phase portrait.

With this background, we consider next examples where initial noise is present. Writing the output time series after n iterates as

$$\hat{v}_n(t) = V(t) + \epsilon\hat{\eta}(t), \qquad (10)$$

the SNR gain, letting $\epsilon\eta(t)$ denote the initial noise, $\|\eta\| = 1$, is

$$\delta_n = \delta(\alpha) = 20\log\frac{\|V\|}{\epsilon\|\hat{\eta}\|} - 20\log\frac{\|V\|}{\epsilon\|\eta\|} = 20\log\frac{1}{\|\hat{\eta}\|}. \qquad (11)$$

For initial SNR, $S_i = 20dB$, the gain, $\delta = \delta(\alpha)$, rises to a maximum, $\delta_M = \delta_M(d)$, after $n_M = n_M(d)$ iterates, and then falls under further application of the LGP procedure (Figure 2(a)). The maximum gain itself, in an analogous way, first rises with d, peaks at $d = d_{pk}$, and then falls with further increase of d (Figure 2(b)). These features are not a peculiarity of the choice of Level 0 or of $f = 0.0$ used for Figure 2, but are generic. The $\delta(\alpha)$ plots are somewhat broader for $f = 0.5$ than for $f = 0.0$, but the peak improvement $\delta_{pk} = 10.1dB$, is the same for both f-values (Figure 3). The main difference between the two cases is in the $\alpha_M(d)$ data: the number of iterates needed for maximum gain, $\delta_M(d)$, is significantly fewer for $f = 0.0$, and $\alpha_{pk} = \alpha_M(d_{pk})$ values are also lower (Figure 4).

When we go to Level 1, the story is exactly the same, except that the peak improvements, $\delta_{pk} = \max_d \delta_M(d)$, are about $1dB$ higher. For zero initial noise, as we have seen, Levels 0 and 1 performed the same, while for $S_i = 20dB$ (10% initial noise) Level 1 thus does only marginally better. We look at $S_i = 0dB$ next.

We fix $f = 0.0$ and compare performance for Levels 0 and 1 with reference time series taken to be $V(t)$ (Figure 5). Here Level 1 outperforms Level 0 by about $6dB$, or, in other words, by a factor of 2. We compare these results with those from a similar study, but with two different reference time series taken from the output of Level 0 computations on the same $S_i = -0.1dB$ time series-corresponding to (1) $d = 25$, $n = n_M = 79$, with $f = 0.5$ and $\Delta = 1$, for which the output $SNR=12.4dB$, and (2) $d = 25$, $n = n_M = 13$, with $f = 0.5$ and $\Delta = 2$,

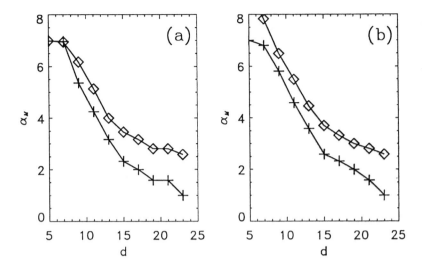

FIG. 4. Log of the number of iterates required to reach maximum SNR improvement for the $20dB$ Lorenz data with $f = 0.0$ (+), $f = 0.5$ (◇). (a) Level 0 and (b) Level 1. Corresponding SNR gain data are shown in Figures 2 and 3.

and for which the output $SNR=12.8dB$ (Figure 6). For case (1), with $d = 25$, δ_M appears to be near its peak value, but is still rising, while for case (2) $d = d_{pk}$. The difference between the performance results for the $S_i = -0.1dB$ noisy Lorenz data portrayed in Figure 5 vs. Figure 6 shows that the gains achieved depend upon the quality of the reference orbit. We note that using either of the cleaned up orbits as reference gives a small but nonnegligible 2–3dB gain. The results of our studies for comparison of performance for Levels 0 and 1 for the noisy Lorenz time series case are summarized in Table I.

Finally, we also have a couple of preliminary results to report for Levels 2

TABLE I. $\delta_{pk}(dB)(d, \alpha_M)$ Lorenz study, with $f = 0.0$. The last two columns ((1), (2)) refer to the results from using the two cleaned up time series (1) and (2) mentioned in the text as reference.

S_i	$20dB$	$0dB$	$0dB(1)$	$0dB(2)$
Level 0	10.1 (17,1.62)	12.0 (25,3.70)	—	—
Level 1	10.8 (11,4.75)	17.9 (23,4.17)	14.0(35,2.58)	15.1(31,2.81)

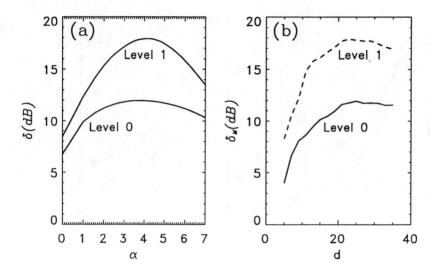

FIG. 5. Level 1 algorithm performance using reference data $V(t)$: (a) $\delta(\alpha)$ for $d = d_{pk} = 25$ (Level 0) and $d = d_{pk} = 23$ (Level 1). (b) $\delta_M(d)$, $d = 5, 7, \ldots, 35$.

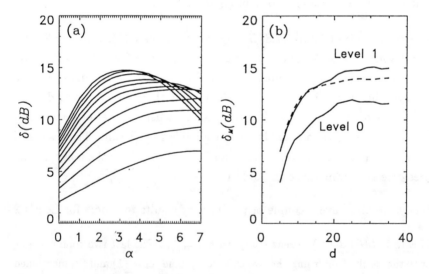

FIG. 6. Level 1 algorithm performance using reference data from Level 0 output time series $\hat{v}(t)$: (a) $\delta(\alpha)$ for $d = 5, 7, \ldots, 25$ and time series (1); (b) $\delta_M(d)$, $d = 5, 7, \ldots, 35$; for Level 1 data — time series (1) (dashed) and time series (2) (solid).

and 3. They are summarized in Table II; the reference time series chosen was the original time series $V(t)$.

TABLE II. Peak SNR gain $\delta_{pk}(dB)$ for $0dB$ Lorenz time series with $f = 0.0$.

	$\Delta = 1$	$\Delta = 2$
Level 2	—	22.3
Level 3	28.5	27.3

SUMMARY AND DISCUSSION

We have presented preliminary results on performance characteristics of the LGP noise reduction algorithm when a reference time series is used to modify the noise reduction process. We have focussed mainly on the dependence upon initial noise strength for the lowest level of use of reference orbit data, in which the only use is for specifying nearest neighbors (Level 1). We have noted that there are two principal kinds of reference data, one where the time series coincides with that of the noise–free part of the given time series, or is close to it, and a second where this may not be the case, but where the underlying attractors are either the same, or are close to one another. In addition, we have described three principal levels of reference time series use for the former case, indicating that similar levels of use can be defined for the latter. In Level 2 the reference data are used not only to locate nearest neighbors, but also the base reference points for the linear affine spaces used to define the noise reduction procedure. In Level 3 the reference data are used to locate the entire spaces, including their orientations in $\mathbb{R}^d$. Finally, we have given a few preliminary results for Levels 2 and 3 use.

In our comparisons we found that Level 1 performance against that of Level 0, which is the nominal (unmodified) LGP procedure, gave no improvement in the absence of noise (baseline noise studies), marginal ($1dB$) peak improvements for 10% initial noise, and significant ($6dB$) gains for the case of 100% initial noise. When LGP noise–reduced time series were used for reference data, the resulting gains were 2–$3dB$.

Our broader goals in these studies are two–fold: (1) to obtain information about the theoretical properties of chaotic noise reduction algorithms, and (2) to devise means to upgrade such algorithms, in particular the LGP algorithm. Hence we seek performance systematics for idealized cases, how these compare with those for non–ideal, noisy cases, and information about algorithm performance limits. There is further work to be done. For instance, how do $B(\alpha)$ plots compare against corresponding $SNR(\alpha)$ plots for various amounts of initial noise? How much, and when does baseline noise limit Level 0 performance? How do these depend upon level of reference orbit use?

ACKNOWLEDGMENTS

This work was supported by ONR and the NSWCDD Independent Research Program.

REFERENCES

1. R. Cawley and G.-H. Hsu, Physical Review A **46**, 3057 (1992).

2. R. Cawley and G.-H. Hsu, Phys. Lett. A **166**, 188 (1992).

3. S. Hammel, Phys. Lett. A **148**, 421 (1990).

4. J. D. Farmer and J. Sidorowich, Physica D **47**, 372 (1991).

5. P. F. Marteau and H. Abarbanel, J. Nonlinear Sci. **1**, 313 (1991).

6. E. Kostelich and J. A. Yorke, Phys. Rev. A **38**, 1649 (1988).

7. T. Schreiber and P. Grassberger, Physics Letters A **160**, 411 (1991).

8. T. Sauer, Physica D **58**, 193 (1992).

9. J. Kadtke and J. Brush, in *Proc. SPIE 1699* (SPIE, Bellingham, WA, 1992), p. 338.

10. P. Grassberger et al., Chaos **3**, 127 (1993).

11. F. Takens, in *Dynamical System and Turbulence, Warwick 1980*, No. 898 in *Lecture Notes in Mathematics* (Springer–Verlag, Berlin, New York, Berlin, 1981), pp. 366–381.

12. T. Sauer, J. A. Yorke, and M. Casdagli, J. Stat. Phys. **65**, 579 (1991).

GLOBAL MODELING OF CHAOTIC TIME SERIES WITH APPLICATIONS TO SIGNAL PROCESSING

J.B. Kadtke
J.S. Brush[*]
Institute for Pure and Applied Physical Sciences (IPAPS)
MS-Q0075, University of California at San Diego, La Jolla, CA 92093

ABSTRACT

Global modeling of chaotic time series involves the extraction of a set of empirical dynamical equations which describe the evolution of the signal in a reconstructed state space. Each equation models the motion of the observed data along a particular empirical degree of freedom, in contrast to the more popular local methods which construct local mapping rules that vary with position on the 'attractor'. Although global methods have presented formidable numerical problems in the past, new approaches have recently been proposed which lead to useful results. In this paper, we present an overview of a basic global modeling scheme, as well as a number of numerical methods to improve the quality and stability of extracted models. We also outline a general approach for the use of global models for signal processing applications, and discuss ways of quantifying model quality and expected signal processing performance. Additionally, we present a number of new noise reduction and signal separation algorithms using global models, and discuss a number of outstanding problems and speculations.

INTRODUCTION

The rapid development of nonlinear dynamics over the last two decades has lead to a new paradigm for our understanding of very complex systems. One of the most natural and powerful applications of these methods has been to time series analysis and, more recently, to related signal processing applications. In recent years, a large number of algorithmic schemes have arisen for the prediction of time series from observed data, and also for noise reduction of observed signals. For certain cases where the observed signals have strong nonlinear components, these new methods have demonstrated significant increases in performance over conventional linear and spectral methods. However, because there is still no unifying approach for

---

[*] at: Rekenthaler Technology Associates (RTA) Corporation, P.O. Box 5267, Springfield, VA 22150

the large number of different analysis schemes, and because quantitative and statistical measures for model assessment are still primitive, the application of nonlinear dynamical and reconstructed state space approaches is still largely an art.

Existing algorithms for modeling and prediction of nonlinear time series can be roughly divided into 'local' and 'global' methods. Currently, the majority of the modeling schemes in use are based on local approximations, and significant results have been reported for a variety of applications[1-8]. These schemes typically involve the explicit reconstruction of the signal as trajectories in a time-lagged embedding space[9,10], whose long term motion is assumed to lie on an attractor (although in practice the reconstructed systems may not possess an attractor in the strict sense). The evolution of the state space trajectories is modeled by defining a simple functional form (or geometric rule) at each point on the attractor, which maps each trajectory forward for a time period for which the local approximation is roughly valid (see e.g. the paper by Cawley, et al. in these Proceedings for more details). These local methods generally have the advantage of being algorithmically simple and easy to implement, are often numerically fast, and are 'data driven' in the sense that there are few arbitrary explicit parameters in the modeling scheme. However, the disadvantages of these methods include a difficulty of rigorous analysis of the scheme due to the often purely algorithmic definition, the introduction of errors due to the piece-wise nature of the evolution rules, the sensitivity to noise of many of the schemes due to the utilization of only limited portions of the state space, and the purely heuristic nature which often leads to specific data dependencies.

An alternative approach to the local methods involves global methods for modeling of the data. Perhaps the first algorithmic schemes for global modeling approaches date to the work of Crutchfield & McNamara[11] and Cremers & Hubler[12]. These and other early methods served to demonstrate the difficulties involved in global approaches. In recent years, however, a number of newer approaches have been presented which have provided significantly improved modeling ability and have addressed some of the fundamental problems[13-16]. As opposed to local methods, global models involve attempting to determine a set of relatively few analytic equations which describe the state space evolution of the data over the observed region of the state space. Each equation describes the data evolution of one empirical degree of freedom, corresponding to one axis of the reconstructed state space, and in essence defines an empirical dynamical system for the particular data set. Such an approach has a number of potential advantages: since we are attempting to find a relatively small number of equations which describe the state space evolution of all the data, there is significant averaging just in the model extraction step which can result in considerable noise reduction. Also, since we generally find an analytic form for the model, it is possible to perform mathematical analyses of the model directly, which can make performance analyses much more direct. In addition, since the models approximate the global evolution, predictive ability for new initial conditions is not limited to those within the actual observed data distribution, and therefore the possibility exists of making predictions in areas of the state space outside of the observed data distribution. Finally, global models offer the possibility of new approaches to noise reduction and signal processing applications, some of which will be discussed later in this paper.

Although global models are potentially very powerful, a number of significant fundamental and numerical problems need to be dealt with. For example, the choice of basis sets for the equation expansions can make a considerable difference in terms of model compactness, however there is no clear way how to determine an optimal representation for a given data set. Also, there is little understanding of the data distribution requirements necessary to determine a relatively unique set of equations, and the related question of sensitivity of the model dynamics to errors in the model coefficients. Because little knowledge exists about these aspects of modelling, determination of a global model can often be a trial and error procedure, with a large number of model choices giving inferior results. Finally, many questions exist as to how to utilize the extracted models in a useful manner for prediction and signal processing applications.

In terms of our approach, we would like to stress that many of the fundamental problems related to global models can by avoided by adopting a more pragmatic philosophy to the modeling of data. If one is concerned largely with only prediction and signal processing applications of observed data, then one can in essence abandon the idea of being able to recover the actual dynamical equations of the physical generating process, which may prove fundamentally impossible in any respect. Our approach is to consider the primary emphasis of global modeling as being the reproduction of the time evolution of the data, firstly from a short term predictive aspect, and secondly for the long term statistics of the signal when possible. By focusing on extraction and refinement of an optimal (in some sense) model for the data, and ways of quantifying the quality of the model, one can define a more reasonable approach for applications.

In the remainder of this paper, we will discuss several aspects of our approach to utilization of global models. These will consist of the general scheme for extracting models, an alternate numerical approach which makes the process simpler, some methods for quantifying the quality of the model and useful parameter ranges, a number of numerical methods for improving the quality of the model, and finally a number of ways to apply the models for prediction, noise reduction, and signal processing applications. The last section will also discuss some remaining problems and some future directions.

EMPIRICAL GLOBAL STATE-SPACE MODELS

In this section we will describe a basic method for the extraction of an empirical global model for a data set. Many of these basic ideas were proposed a number of years ago, although we will develop a number of modifications[11,14]. For our purposes, we will assume that we have constructed a set of D-vectors from the observed data, the ith of which is given by:

$$\underline{Y}[i] = (y_1[i], y_2[i], ..., y_D[i]) \quad (1)$$

These vectors evolve in a D-dimensional state space. We will also assume that we can construct these vectors either from vector time series, or by time-delay (or some other) reconstruction from a single scalar time series, and we will not discuss the details of this procedure here. We note that D will be left as a model parameter which will be varied during the model selection procedure.

We will now outline the procedures to find a global model for the evolution of the above data vectors as either a map, or a set of ordinary differential equations (ODEs) defining a flow in the state space. We note that either choice implies no fundamental statements about the character of the underlying generating process. The choice of the model is somewhat left to the modeler, although coarsely sampled data will generally be more amenable to maps, whereas flows typically require relatively finely sampled data. To describe the basic procedure, we firstly develop the method for modeling the data as a map. To do so, we attempt to find a set of D functions of the state space vectors such that

$$\underline{Y}[i+1] = \underline{F}(\underline{Y}[i]) \quad (2)$$

is a 'good' model of the data evolution, over the entire range of observed data. To determine these functions, we first assume a general form for the model; the most straightforward choice is to assume a polynomial expansion in the vectors $\underline{Y}[i]$:

$$\begin{aligned} y_d[i+1] &= f_d(\underline{Y}[i]) \\ &= a_{d,1} + a_{d,2} y_1[i] + a_{d,3} y_2[i] + ... \\ &= \sum_m^M a_{d,m} b_m(\underline{Y}[i]) \end{aligned} \quad (3)$$

where the b_m define the 'basis set' for the expansion and the $a_{d,m}$ are the coefficients of these bases for the dth component of $\underline{F}$. For polynomials, the order P of the expansion will also be left as a variable model parameter, and the number of terms M in the general expansion will be $(P+D)!/(P!D!)$ for a D-dimensional vector space.

Given this representation, the functions defining the global model can then be written as a matrix equation

$$\underline{Y}_d[i+1] = \underline{B}[i]\underline{A}_d \qquad (4)$$

or in matrix element form

$$\begin{pmatrix} \cdot \\ \cdot \\ y_d[n+2] \\ y_d[n+3] \\ y_d[n+4] \\ \cdot \end{pmatrix} = \begin{pmatrix} \cdot & & & & & \\ 1 & y_1[n+1] & y_2[n+1] & \ldots & y_d[n+1]^P & \ldots \\ 1 & y_1[n+2] & y_2[n+2] & \ldots & y_d[n+2]^P & \ldots \\ 1 & y_1[n+3] & y_2[n+3] & \ldots & y_d[n+3]^P & \ldots \\ \cdot & & & & & \end{pmatrix} \begin{pmatrix} a_{d,1} \\ a_{d,2} \\ a_{d,3} \\ \cdot \end{pmatrix} \qquad (5)$$

where $\underline{B}$ is the data matrix, $\underline{Y}_d[i+1]$ is the vector of values of the dth data component at the $(i+1)^{th}$ data point, and $\underline{A}_d$ is the vector of coefficients to be determined. It should be noted here that although the global model is nonlinear, the nonlinearities have been absorbed in the structuring of the data matrix, so that the equation is linear with respect to the unknown coefficients. It is possible, for more complicated basis expansions, to have terms which are nonlinear in the unknown coefficients. In these cases, the equations are solved by iterative nonlinear least-squares numerical algorithms.

The linear matrix equation above can be solved in a least-squares sense to determine the model coefficients. We have found that the best solution scheme is generally singular value decomposition (SVD)[17], for a variety of reasons relating to numerical stability, efficiency, and avoidance of problems due to matrix singularities. Use of SVD also leads to considerably more flexibility in terms of model refinement, which will be discussed in a later section.

As an example of the above method, we consider an idealized time series generated from the Henon system, a chaotic map defined by

$$x[t+1] = 1 + y[t] - 1.4x[t]^2$$
$$y[t+1] = 0.3x[t] \qquad (6)$$

For this example, data was generated by iterating the Henon system and recording a scalar time series from the x coordinate. Time delay reconstructed vectors were then generated from this time series using a delay of 1, and the matrix equation for the global model was constructed using a general model with $D=2$ and $P=2$. The equation was then solved via SVD to yield the global equations defining the model. If we first note that in time delay form, the Henon equations can be written (found by direct substitution):

$$x[t+1] = 1 + 0.3x[t-1] - 1.4x[t]^2 \tag{7}$$

we then present the results of determining the global model from the reconstructed data:

$$\begin{aligned}x[t+1] = &\ 1 + 0.3x[t-1] + 6.94\times10^{-7}x[t] \\ &+ 4.69\times10^{-6}x[t-1]^2 + 8.83\times10^{-7}x[t]x[t-1] - 1.4x[t]^2\end{aligned} \tag{8}$$

where only 25 data points were used to determine the equations! Although it is perhaps not surprising that good estimates of the original equations can be determined given that the correct order and dimension for the model were known *a priori*, and also that noise-free data was used, the result demonstrates some interesting points. Firstly, for perfect noise-free data and the correct basis expansion, it is in theory possible to determine the N coefficients of any model using only N data points, since we are solving a linear system. In practice, proper sampling of the attractor and numerical issues have some effect, however in this example 25 data points were more than sufficient to obtain excellent accuracy. Secondly, although the correct coefficients were obtained for the original model terms to high accuracy, there generally are a significant number of spurious terms with non-zero coefficients, which typically arise due to solution degeneracies as well as simple numerical errors. Although in this example the coefficients of the spurious terms are quite small, in general these terms can be of comparable magnitude to actual terms, and are generally undesirable due to the numerical errors which they induce. The reduction or elimination of these spurious coefficients is part of what is termed 'model refinement', which will be discussed in the next section.

An alternate class of global models is the extraction of a set of ODEs which model the data evolution in the state space. The choice of this class is often motivated by the fact that most natural deterministic physical systems obey dynamical equations of this form, and so it is a natural and powerful way to attempt to model the signal evolution from data derived from physical generators. The ODE formulation describes an analytic flow of trajectories in a vector space of the reconstructed state space, which can provide a potentially powerful advantage over heuristic local fitting methods. To construct a global model using the ODE formulation, we start from the definition of the equations:

$$\frac{dy_d[i]}{dt} = f_d(\underline{Y}[i]) \qquad (9)$$

which states that we are determining functions of the coordinates (as for the mapping case) which estimate the derivative of the state space trajectory at the same data point. Because the form of the equations specify the derivative of the state space trajectory at each point, the resulting state space trajectories are required to be smooth and continuous. If we construct a matrix equation using the data matrix and coefficient vector, as for the global maps, then this system can be solved by requiring that

$$\underline{Y}[i+1] = \int_i^{i+1} \underline{B}[i] \underline{A}_d \, dt \qquad (10)$$

or rather that the integral of the model using trial model coefficients starting at the ith data point must result in the $(i+1)^{th}$ data point. Determination of the model coefficients in this manner requires an iterative search scheme, which is typically numerically intensive, rather inaccurate, and difficult to generalize.

A more direct approach is to attempt to approximate the derivatives on the left side of Eq.(9) directly from the data[14]. Using this formulation, a matrix equation can be constructed exactly as Eqs.(4) and (5) for the global mapping case, with the left side being filled with the local derivative approximations, and the problem once again reduces to a linear matrix equation which can be solved via SVD. We note that although the numerical difficulties have in part been reduced, some of them are now transferred to the problem of accurate estimation of the trajectory derivatives. For significantly noisy data in particular, this can be a difficult problem, and the error in slope approximation can often be a limiting source of error in model extraction. Several numerical schemes can be utilized, however, to significantly improve the accuracy of these estimations[18].

The advantages of solution by direct derivative approximation are many. Beside the numerical efficiency and speed gained by reduction to a linear matrix equation, this formulation allows for generalizations of the differential flow approach to other signal processing applications such as adaptive modeling, which will be discussed in the accompanying article in these Proceedings[19].

As an example of estimating system coefficients from a set of ODEs, we consider the Lorenz system. This system is an idealized chaotic dynamical system of ODEs defined by

$$dx/dt = 16(y-x)$$
$$dy/dt = 45.92x - xz - y \quad (11)$$
$$dz/dt = xy - 4z$$

The original system above is a second order system with three degrees of freedom. As an example of the ability to recover the dynamical equations for a differential flow, we extract vector data consisting of time series from the x, y, and z coordinates and construct the matrix equation as per Eq.(5). Note that here, we will use vector data so that the reconstructed equations can be compared to the original system by the reader. Time delay reconstruction for flows generally results in extracted equations which are not directly recognizable, because of the nonlinear coordinate transformation which the embedding induces. Although predictive ability for time-delay reconstructed systems are reasonably comparable to the vector reconstruction versions, it is more difficult to directly compare the system to the original, and can usually only be done via indirect measures of predictive power.

Using 200 points with a sampling interval of 0.005 from Eqs.(11), the Lorenz equations were reconstructed as shown below (boldface is used to highlight the terms which appear in the original equations):

$$\begin{aligned}
dx/dt = &\ -0.2269 - \mathbf{16.11}x + \mathbf{16.08}y + 0.0173z \\
&+ 0.00072x^2 + 0.00077xy + 0.0025xz \\
&- 0.00088y^2 - 0.0021yz - 0.00030z^2 \\
\\
dy/dt = &\ -0.6839 + \mathbf{45.58}x - 0.8694y + 0.0423z \\
&+ 0.0015x^2 - 0.0012xy - \mathbf{0.9940}xz \\
&- 0.00017y^2 - 0.0018yz - 0.00058z^2 \quad (12)\\
\\
dz/dt = &\ -1.255 + 0.0076x + 0.0056y - \mathbf{3.871}z \\
&+ 0.0194x^2 + \mathbf{0.9783}xy - 0.00019xz \\
&- 0.0053y^2 - 0.00003yz - 0.0026z^2
\end{aligned}$$

As can be seen, the terms appearing in the original system have reconstructed coefficients very close to the original values, however as before there are spurious terms appearing with coefficients of non-negligible size. Also, note that only 200 data points were used in the construction, and considerably fewer points could have been used at the expense of only a small amount of accuracy.

To determine the quality of the extracted model, it is instructive to investigate the ability of the model to predict the original time series. In Figure 1, we show a segment of the original time series of the x-coordinate of the data, and a prediction of that segment using Eqs.(12), starting from the first point of the segment. As can be seen, the predicted trajectory shadows the original trajectory very closely as far as seven characteristic orbital periods. Since in general signal processing applications the form of the underlying model is unknown, we assume that the primary criterion for judging quality of the extracted model must be the ability to predict, i.e. to regenerate the underlying time series (minus the stochastic components). Use of this model criterion, as well as ways of quantifying this predictive level, is critical to our signal processing approach, and we will discuss this in some detail later in this paper.

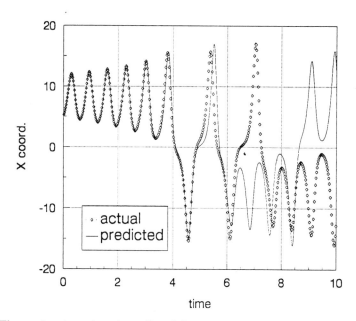

Figure 1. Actual and predicted Lorenz x coordinate from Eqs.(12), derived from 200 noise free data points.

The procedure outlined above is a very simple and basic method for the extraction of a global model for a time series. In practice, though, the straightforward application of this method is too unsophisticated to attain reasonable results in practical, real-world signal processing applications. The remainder of this paper will discuss a variety of ways to improve this modeling approach. To begin, we list a number of general properties of the above model extraction procedure:

> As mentioned, for very clean (i.e. relatively noise-free) data, one can often obtain good model estimates using a relatively small number of data points. An important point, however, is that the data points used sample reasonably well the majority of the attractor structure, or in other words evenly sample the region of the state space visited by the

data. Although this is hard to directly quantify, a good rule of thumb is that sparsely distributed points should be taken at least one characteristic signal period apart.

- Errors in local derivative approximations can be significant and can often be a limiting source of error. Since, for noise free data, the error in the approximation scales with the sampling interval size, it is usually advantageous to have data samples relatively closely spaced together where slopes will be calculated. This fact, combined with the previous observation, suggests that a very efficient means of data sampling would be 'burst' sampling, i.e. sampling sets of three to five data points at high sampling interval, but recording these sets relatively infrequently.

- The presence of noise in the data generally increases the errors in the extracted model coefficients. However, we find that increasing the number N of data points used in the construction of the matrix equations decreases the error in the resulting model by roughly $N^{1/2}$.

- A surprising qualitative point is that one can often get sufficiently good models using only low-order polynomial expansions for the global models. Although these expansions may not properly describe the underlying dynamical system, they often provide sufficiently good local predictive power to perform a variety of necessary signal processing applications.

- The presence in most derived models of 'spurious' dynamical terms with non-zero coefficients generally introduces significant numerical noise and un-necessary computational load. The reduction of these spurious terms is thus a reasonable goal, and general conceptual notions of 'model entropy' (such as the minimum description length principal from information theory) have been discussed in detail elsewhere. We will therefore discuss a number of numerical methods for the reduction and elimination of these terms in later sections.

- Finally, we will show later that the particular approach of global ODE model extraction by direct derivative approximation lends itself readily to a variety of other signal processing applications, such as real-time adaptive methods and detection/classification schemes which utilize nonlinear dynamical signal content.

QUANTIFYING MODEL QUALITY

In order to effectively apply a modeling approach to applications such as signal processing and noise reduction, it is imperative that tools exist to determine the 'goodness' of a given extracted model in reproducing the signal components of interest in a data set. The generic situation in signal processing or time series problems is that we have little or no knowledge of the detailed structure of the underlying system, hence we cannot choose *a priori* an appropriate model dimension or order, or even a model class, as in the examples of the previous section. This leads to the question of what measures and criteria one should use to quantify the quality of a given model. For a general situation, a powerful choice that we utilize is the criterion of maximizing the predictive ability of the model with respect to the original data, or in other words the model should optimally reproduce the characteristics of the signal. Generally then, we choose as our modelling philosophy the criterion that 'best model is the best predictor'. For simple signals consisting of one deterministic component (plus possibly noise) this concept is straightforward; for more complicated signals consisting of several components, the problem becomes one of signal separation, which will be discussed elsewhere.

The necessity of being able to diagnose model quality for the global modeling procedure outlined above can be understood in the following intuitive terms: for both the map and ODE model classes, we fit our model subject to the constraint that the one-step predictions or slope estimates are best fits to the observed data. Clearly, this leaves some degeneracy in the solution sets, since we do not provide constraints for the n-step predictions of the model (i.e. there may be an entire class of models that roughly equally well reproduce the one step predictive characteristics). The addition of noise in the data further complicates this degeneracy problem, and introduces errors in the coefficients in any given model realization. Finally, since the SVD numerical solution technique for the matrix equation chooses a minimal norm solution (as do most others)[17], a model realization may be selected by SVD which is perfectly valid given the one-step prediction constraints, but which may vary considerably in terms of optimal long term predictive power.

As a consequence of the above remarks, we follow a general procedure whereby we extract, for any given data set, a variety of models of different order and dimension (and possibly basis representation) and evaluate each according to a number of different model quality diagnostics. The two primary diagnostics are listed below:

- RMS residuals of the least-squares SVD fit: As a crude estimate of the model fit, the residual of the least-squares fit can be monitored during coefficient fitting. For normalized residuals which are poor, the model is in no sense a good representation of the observed data, but the input noise level must be taken into account to interpret the numerical values.

- Dynamic correlation: The dynamic correlation measure (D[s]) of Kravtsov[20] is a normalized measure of the RMS error between the s-step predicted trajectory segments and the actual data, as a function of number of prediction steps forward s. This measure has a useful normalization and is numerically efficient. Again, the crucial element here is that we measure predictive ability as a function of prediction time s, which allows us to gauge the useful predictive time scales for applications.

As an example of the dynamic correlation, Figure 2 shows a typical D[s] curve generated for a model extracted from Henon data. As can be seen, the model provides near-perfect predictive power on average (i.e. D[s] values near unity) over a significant range of s-step predictions. After about 20 iterations, the average predictions become poor very rapidly, a consequence of the exponential divergence of state space trajectories for chaotic systems.

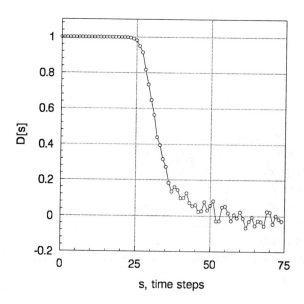

Figure 2. Sample D[s] curve derived from model of the Henon system.

The principal point here is that general measures of maximal theoretical predictive time scales for a given system (e.g. the largest positive Lyapunov exponent) do not give useful information about particular data sets and particular extracted models. In general, we do not expect to be able to determine models which give the theoretically maximal prediction times in practice. A measure of, e.g. the D[s] curve for a specific model does however give information that is specific both to a particular data set and to a particular derived model. For a given model, it gives vital information as to the range of time scales over which the model is a valid predictor. The size of this range is important and useful firstly as a measure of the quality of a given model, but also secondly to determine the 'operating points' on the predictive curve that can be used to choose parameters for, for example, a global model-based noise reduction scheme. Therefore, one typically routinely generates a D[s] type measure to aid in the model selection process.

Besides the measurement of predictive ability of given models, which is essentially a local measure, it can also be important to assess the global characteristics of a given model. By global, we mean the ability of the model to reproduce the long time statistical properties of the original signal. This measure is motivated by the fact that it is possible to extract global models with relatively good local predictive ability, but which produce erroneous signal statistics when given initial conditions are evolved for long periods of time under the model dynamics. For example, an extracted model for the Lorenz system may produce long term trajectories which are quasi-periodic or even periodic rather than chaotic for long time evolutions. Although such behavior may often have no bearing on the particular signal processing application one is interested in, from a pure modeling or physics standpoint it can be important to examine such model characteristics. Currently, however, few efficient measures of such long time statistical behavior are in use, and one typically simply iterates the model and examines the resulting global state space structure manually.

REFINING THE GLOBAL MODELS

As mentioned above, because of the constraints used to derive the global models using the basic techniques, the actual model realization obtained may be only one of a set of models which satisfy the one-step prediction criterion. This set of models may vary considerably in terms of multi-step predictive ability, global statistical properties, and model entropy. In this section, we will now discuss a few numerical methods which can be used to refine the particular model during the fitting procedures, which can increase the ability of the resulting model to reproduce the data characteristics[21].

Because of the simple linear form of the matrix equation using the data matrix approach, one can perform several manipulations during the fitting procedure to reduce the effects of random noise components in the data, as well as reduce the number of spurious model coefficients and variables. We briefly outline the manipulations below.

Generally, the criterion of choosing a minimal norm solution by the least-squares procedures means that the variance of the model coefficients will tend to be 'spread' across the available basis set terms of the trial model, including terms which are spurious to the actual data. Particularly, the resulting model will tend to retain terms up to the highest order of the trial model, whether these are actually represented in the data or not. One way of minimizing this effect is by rescaling of the data matrix prior to least-squares solution. This method consists of rescaling all n^{th} order variables x^n to $\alpha^n x^n$, by some constant $0.0 < \alpha \leq 1.0$. Although this re-scaling does not change the value of the residual of the fit (and hence the one-step predictive ability of the model realization), it does force the solution vector to represent the data using the lower-order terms of the trial model (since coefficients for the reduced terms must be relatively large, violating the minimum norm requirement).

In a manner analogous to that described above, one can rescale the number of variables in the trial model, i.e. the dimensionality of the state space representation. In this way, the solution vector will be forced to represent the data using the fewest possible number of independent model variables. The effect of both of these rescaling schemes is to therefore reduce the number and variance of spurious terms in the model, thereby reducing model entropy and the resulting numerical noise and inefficiency.

A third technique utilizes the singular value spectrum resulting after the SVD decomposition itself. Editing and truncation of the singular value spectrum in signal processing applications is not new, and can be utilized here to reduce the random components in the model representation as well. Typically, one can begin setting singular values to zero starting with the smallest in magnitude, and continue the procedure until there is a significant change in the least-squares fit residue, at which point one returns to the previous solution. More sophisticated methods for choosing the cutoff of the singular value spectrum utilizing information theoretic methods are currently being investigated by the authors.

One particularly powerful method that we have utilized for reducing model entropy is a 'zero-and re-fit' procedure for the model coefficients. This procedure consists of two simple steps: firstly, the magnitude of the coefficients of a particular extracted model are examined, and a cutoff value is chosen (typically 10^{-3} or 10^{-4} of the largest model coefficient). All coefficients smaller in magnitude than this value are then set to zero. This step is tantamount to an *ad hoc* procedure for identifying and eliminating spurious terms of the extracted model which are small in magnitude. This procedure is not equivalent to improving the model quality, however, since we have generally made the predictive ability of the model worse by removing terms. Therefore, we define a second step in the procedure: the model is re-fit to the data, now omitting the basis terms corresponding to the coefficients which have been zeroed. This step therefore recovers the predictive ability of the given model. It has been pointed out to us recently that a similar general procedure is utilized in the field of statistics and is known as 'confirmatory factor analysis', and has no doubt been utilized elsewhere.

In practice, after the refit step above it occasionally occurs that other terms of the model may fall below the cutoff magnitude; hence the procedure can be applied repeatedly until the model coefficients stabilize. We generally find in tests on idealized systems that the majority of spurious terms are driven to zero after two or three cycles of the iteration procedure, and the relevant model terms stabilize to near their correct value. This procedure can therefore prove quite powerful in model refinement, and is particularly useful for noisy data.

As an example of model refinement, we consider a global flow model derived from 10,000 points of noisy vector data of the Lorenz system at a simulated signal-to-noise ratio (SNR) of 25 dB. Eqs.(13) show the derived model coefficients firstly using only the straightforward extraction procedure (boldface is used here and elsewhere to indicate which terms are in the original system).

$$\begin{aligned}
dx/dt = &\ -2.055 - \mathbf{14.25}x + \mathbf{14.98}y + 0.1770z + \\
&\ 0.01510x^2 - 0.01312xy - 0.02715xz + \\
&\ 0.00219y^2 + 0.01352yz - 0.00310z^2
\end{aligned}$$

$$\begin{aligned}
dy/dt = &\ 5.215 + \mathbf{44.52}x - \mathbf{0.2549}y - 0.1489z - \\
&\ 0.00385x^2 + 0.00342xy - \mathbf{0.9788}xz - \\
&\ .00101y^2 - 0.01132yz + 0.00089z^2
\end{aligned} \quad (13)$$

$$\begin{aligned}
dz/dt = &\ 1.4829 + 0.02786x + 0.1255y - \mathbf{4.113}z + \\
&\ 0.00354x^2 + \mathbf{0.9837}xy + 0.00144xz + \\
&\ 0.00764y^2 - 0.00051yz + 0.00171z^2
\end{aligned}$$

Eqs.(14) show the resulting coefficients after all four of the procedures outlined above have been performed.

$$dx/dt = -\mathbf{15.85}x + \mathbf{15.86}y$$

$$dy/dt = \mathbf{43.14}x - \mathbf{0.9753}y - \mathbf{0.9642}xz \quad (14)$$

$$\begin{aligned}
dz/dt = &\ -0.08938 + 0.08687x + 0.09451y - \mathbf{3.990}z \\
&\ + \mathbf{0.9977}xy
\end{aligned}$$

As can be seen, there is significant reduction in the number of spurious terms, and the relevant terms are significantly closer to the idealized Lorenz equations.

One important effect of model refinement is that these procedures can often significantly improve the ability of the model to recover the long-time statistical properties of original data set. For example, if one monitors the dynamical invariants of data generated by the refined models, such as Lyapunov spectrum or the attractor dimension, one generally finds a significant improvement in these dynamical measures[21]. The refined models are therefore 'closer' dynamically to the original generating system, and therefore may be superior models if one is concerned with statistical probabilities of the data. As a related point, since calculation of dynamical invariants are often very difficult for significantly noisy data (particularly for Lyapunov exponents, calculations of which typically break down at noise levels of 35-40 dB), an alternate approach may be to first extract and refine a global model from the noisy data, utilizing the noise averaging characteristics of the modeling procedure. The dynamical invariants can then be calculated from clean time series obtained by integrating the global model, thereby bypassing many of the difficulties related to local approximation of the flows for noisy data. Previous experiments have indicated that this approach often produces good estimates for the Lyapunov spectrum even from data with noise levels down to 20-25 dB.

APPLICATIONS

In the previous sections, we outlined some of the basic methods for extracting and refining global state space models for time series of data. In the remaining section of the paper, we will present a number of new ways which the extracted models can be utilized for signal processing and classification/detection applications. The applications of nonlinear dynamical modeling to time series analysis can be placed into several categories, such as prediction, noise reduction, and signal classification. Although at least the first two of these have received considerable attention in recent years, very little has been done in the way of utilization of global modeling methods. Because of the form of the global models, new ways of approaching the above applications can be defined that differ from the local approximations currently employed. In the remaining part of this paper, we will briefly outline several new methods which we have developed for these signal processing applications.

APPLICATION: PREDICTION

Since much of the previous discussion has revolved around measuring and improving the predictive ability of extracted global models, we will not discuss these issues to any length here. We will however stress two further points: firstly, since global models are fit to whole regions of the state space in the vicinity of the data, their functional forms (by analytic continuation) can be quite useful for predicting initial conditions even where the original data was absent or sparsely populated. This is in contrast to most local approximation schemes which require at least a reasonable population of observed data to extract sufficient information about the local state space dynamics to perform prediction. Currently, however, the extent to which global models can be used to predict in regions away from observed data, which is fundamentally a question of how uniquely a given amount of data recovers the underlying generating system of the data, is still only poorly known.

Secondly, even for good global models, it typically occurs that there are regions of the state space for which the model is a much poorer predictor than on average. Thus, for applications where prediction of near-future values of the time series is crucial, an important tool can be the construction of a measure of the models predictive ability as a function of attractor position. This can easily be crudely constructed by covering the whole region of interest of the state space with initial conditions and evolving them using the model, and defining some measure of the predictive ability which is then charted along the extent of the attractor. The point of this is to identify regions of the state space for which the model may give poor or erroneous predictions, so that initial conditions starting in these regions can then later be treated with appropriate suspicion.

APPLICATION: NOISE REDUCTION

Noise reduction is an important application of nonlinear dynamical time series analysis, and has received considerable attention in recent years. The ability to extract good global models implies that it may be possible to define new classes of noise reduction methods. Below we will outline two simple ways to perform noise reduction which utilize the specific character of global models[22]; more details of these methods can be found in the references.

The first scheme is a simple modification of a noise reduction method which uses local geometric information to smooth a noisy trajectory point in the state space of the observed time series. The method on which it is based can be considered a generic class which is similar to the recent Grassberger-Schreiber geometric method[8], and also similar to the Farmer-Sidorowich method[1]. The basic local scheme is to choose a noisy data point of the observed data in the reconstructed state space, and to determine neighboring points on nearby trajectories of the observed data out to some specified radius. This group of neighbors is then followed forward in their time evolution, and the smoothed version of the evolved noisy data point is then taken as some averaging function (e.g. centroid) of the evolved positions. One difficulty with these particular schemes is that they are sensitive to the distribution and density of neighbors for any particular prediction, therefore often making data requirements quite high.

As an alternate method, we assume that we can extract a good global model for the given set of observed data. We then note that the use of the positions of actual data points as neighbors is superfluous. Since we assume we have a good representation of the actual dynamics over most places of the attractor, we can generate arbitrary numbers of 'artificial' neighbors (i.e. initial conditions) at any desired distribution around the noisy point to be cleaned. The extracted global model is then used to evolve the artificial neighbors and perform smoothing as per the original scheme. The improvement in performance is obtained because of the large number of neighbors which can be evolved and averaged.

In practice, this scheme provides reasonable improvement over the basic scheme, which improves as the number of neighbors at each point is increased. However, the computational load can also increase significantly with increasing numbers of generated neighbors, and the results are sometime sensitive to regions of poor predictive ability of the global model.

We have generally found much better results using another alternate scheme, which is a prediction based noise reduction method. This method, which we term 'evolved trajectory minimization' (ETM), is described as follows: let us again assume that we have extracted a good global model for the dynamics, which is on the average a good predictor of the data out to s time steps. We now consider a noisy observed data point on a trajectory in the reconstructed state space. We can utilize the extracted global model to evolve the noisy data point forward a series of s steps, and compare to the actual noisy trajectory segment. If the system is chaotic and nearby trajectories separate exponentially, then the trajectory segment evolved from the noisy initial condition should rapidly diverge from the original noisy trajectory (although it is obviously not a requirement that the data be chaotic). If we construct a cost function which measures the least-mean-square distance between the s-step predicted trajectory and the original trajectory, then we can try to minimize this distance by varying appropriate parameters. In this case, the parameter to be varied is the initial location of the noisy data point. Thus, we move the data point around in state space until its s-step predicted trajectory is as consistent with the original noisy trajectory as possible. In practice, the minimization of the cost function can be trivially performed with a multi-dimensional gradient search routine[17].

We have found the above method to be quite powerful even when global model of only modest predictive ability can be extracted[22]. We summarize some of its important properties below:

- If the given global model is a good predictor out to s time steps, then the increase in SNR of the signal in one pass of the method is proportional to $s^{1/2}$.

- If the noise is zero-mean and the global model is perfect, then convergence to the original clean data trajectory is guaranteed.

- A 25 to 30 dB improvement in SNR has been observed in a single pass. Performance is however dependent on sampling rate, predictive power of model, and value of D[s] for the s-step prediction as well.

- Multiple passes are possible and generally suggested.

- The method has proven useful for initial SNRs down to 0 dB.

As an example of the method, we consider a time series from the Lorenz system with Gaussian distributed random noise components to the time series to simulate an SNR of 0 dB (Figure 3). Using the true Lorenz equations and projecting ahead about one orbit for the minimization results in a SNR increase of about 32 dB. Figure 4 shows the recovered Lorenz attractor after one pass of the ETM method. Details of the technique, along with effects of sample sizes and noise, can be found in Kadtke & Brush[22].

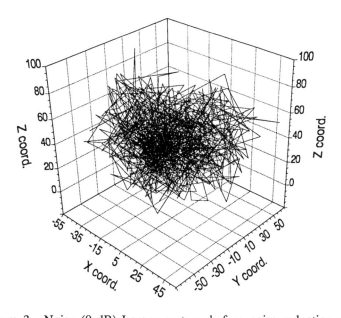

Figure 3. Noisy (0 dB) Lorenz system, before noise reduction with ETM method.

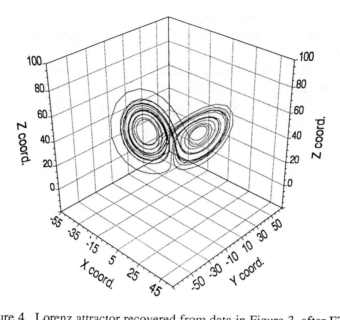

Figure 4. Lorenz attractor recovered from data in Figure 3, after ETM noise reduction.

One limitation of the ETM method is that the maximum SNR that any noisy time series may be improved is dependent on the quality of the extracted global model. Intuitively, this occurs because any imperfect model itself introduces some amount of error during its application, which limits the resulting noise level after noise reduction method. This general idea is in keeping with recent results reported by Cawley, et al. elsewhere in these Proceedings. We have been able to derive a simple expression which relates the maximum amount of increase in SNR possible for a given data set and a given model, regardless of the noise reduction method used, as a function of the $D[s]$ curve for the model[23]. This relationship is given by

$$\log(RMS_s^2) = \log(1 - D[s]) + \log(2\sigma^2) \qquad (15)$$

where the left side represents the log of the squared RMS error between the original and s-step predictions of the data and σ is the standard deviation of the original data set. For a particular model operating at a value on the $D[s]$ curve defined by an s-step prediction, this relationship gives the maximum amount of noise reduction possible using this particular model.

APPLICATION: 'RARE EVENT' MODELLING

As an interesting aspect of the potential of global modeling approaches, we consider the possibility of prediction into regions of the state space which have not been visited by the observed data. Although intuitively the possibility of prediction away from the observed region seems very limited, the issues involved in global modeling are tied to ones of model uniqueness and dependent on data distribution, and are not well understood. The potential implications are obvious, but in particular may be important to the modeling of 'rare' or 'extreme' events. These events, defined as events which occur very infrequently but may have catastrophic effects (e.g. extreme meteorological conditions), may be identified with very low probability regions of an attractor in the state space of a dynamical system. One obvious question is whether it is at all possible to predict such events if one has not been observed during the period of data observation. To give some intuition to this question, we perform a simple numerical experiment: we record a time series from a variable of the Lorenz system, but one which is confined only to observations lying on a small region of the Lorenz attractor (Figure 5).

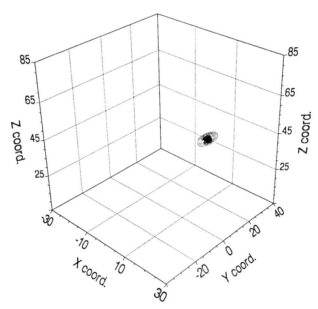

Figure 5. Small portion of Lorenz attractor used to learn system dynamics.

We then use this data to reconstruct equations of motion for the system. Using this extracted global model, we regenerate a time series to determine the long term distribution properties of the state space trajectories. This trajectory is shown in Figure 6, and clearly regenerates the full structure of the Lorenz attractor to reasonable accuracy. This experiment and other similar ones therefore seem to hint that global models may at least in principal prove to be useful for predicting into regions of the phase for which one has yet to observe data experimentally.

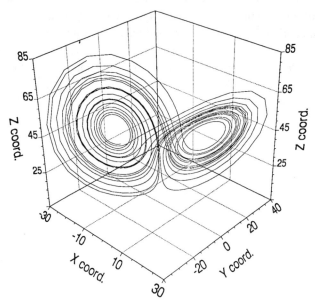

Figure 6. Attractor generated from equations derived from data in Figure 5.

APPLICATION: DETECTION & CLASSIFICATION OF SIGNALS

One of the most important potential applications of nonlinear time series analysis methods may be to the detection and classification of signals. Nonlinear time series methods are structured to utilize the nonlinear information inherent in any given signal, to which conventional linear methods (i.e. spectral or correlation techniques) are insensitive. Since the majority of signals with deterministic components derived from physical processes contain some nonlinearity, considerable potential exists for the exploitation of this information using these new techniques.

In the reconstructed state space framework which nonlinear dynamical methods utilize for signal analysis, nonlinearities in the deterministic signal components appear as some phase couplings in the nonlinear modes of the generator. This information generally dictates the topology of the object on which the long time signal evolution lies. A natural way to attempt a detection/classification scheme in the state space is therefore to use some method of quantifying this signal topology. In this respect, extraction and comparison of global models can be quite useful since they are naturally sensitive to the underlying global state space structure which the signal generates, and provide a compact form for its description.

We have defined and investigated a number of new methods for the detection and classification of signals which utilize global models as the fundamental state space measure. Generally, we have found discrimination abilities that are quite significant, can detect the presence not only of deterministic components but components from different generating systems, and can do so even to noise levels below 0 dB. These particular techniques and results will be discussed to some detail in the accompanying article in these Proceedings[19].

APPLICATION: ADAPTIVE GLOBAL MODELING METHODS

A final application which we will mention here is a formulation for extraction of global models which is based on a formulation for conventional adaptive linear filters[19]. Because of the form of our global modeling scheme, even the ODE formulation can be readily modified so that the model coefficients are calculated by a continuous updating scheme. In this way, model coefficients converge to the correct coefficients for the data in a smooth and continuous fashion, and can track changes in model coefficients caused by a possibly time-dependent physical generating system. The significance of this development is that these adaptive, nonlinear state space modeling schemes are now in principal capable of modeling non-stationary as well as nonlinear data. Although this general scheme has been developed only recently, it has already proven to be quite powerful and to be applicable to a number of novel detection/classification schemes, for which conventional linear adaptive filters and stationary nonlinear dynamical methods are insufficient. The bulk of the accompanying article in this Proceedings is devoted to a discussion of these adaptive techniques.

CONCLUSIONS AND OPEN QUESTIONS

Although the development of new global modeling techniques in recent years has made them a potentially powerful tool, there are a large number of open questions and unknown aspects still to be explored. We list some of these major points below:

New Basis Sets

An important and outstanding question is how to define and utilize new basis sets for the general model expansions of the evolution functions. More important than this, however, may be the question of how to *a priori* choose an optimal basis representation based on the signal characteristics. Very little is understood about this problem currently.

Data Requirements

Particularly for applications to detection/classification, an important and fundamental question is how much data (in the sense of populating the state space) is necessary to uniquely define a given model for the data. Although often relatively small amounts of data are necessary to recover models in low noise cases, no clear guidelines exist for how to approach this issue.

Noise Reduction

Clearly the form of global models allows for new ideas as to their utilization in noise reduction schemes. Although we have suggested two here, a wide variety of possibilities exist.

Detection/Classification

The usefulness of global models for detection and classification of signals based on nonlinear information may be particularly apt. As with the noise reduction application, however, possible schemes for their utilization are only beginning to be explored.

Adaptive Modeling

The development of adaptive modeling schemes make it possible to deal with signals which are both non-stationary as well as nonlinear, and hence make it possible to model signals of entirely new classes. The adaptive formulation may also make it possible to implement these modeling algorithms in a real-time framework, possibly on an inexpensive single signal processing chip.

Physical Information

Although it is in general not possible to recover the original dynamical equations of a physical generating process, it may be possible to compare the structure of postulated physical models, under the reconstructed state space transformation, to that of extracted data. Because global models (i.e. the ODE formulation) are of the same general form as dynamical systems, they may be more appropriate than other model classes for this. Although no clear method for this has been demonstrated, it seems at least feasible that in this way some physical information about the underlying system can be deduced.

ACKNOWLEDGEMENTS

The authors would like to acknowledge useful discussions with Dr. Joachim Holzfuss of the *Institute fur Angewandte Physics* at the University of Darmstadt (Germany), who contributed ideas to some of the above research, and JBK would like to acknowledge support by the NATO Office of Scientific Affairs for the research travel grant #900088 which allowed these discussions to take place. Both JBK and JSB have been supported in large part by ONR contract #N00014-92-C-0045, DARPA contract #DAAH01-92-C-R197, and RTA IR&D funding.

REFERENCES

1. J.D. Farmer and J.J. Sidorowich, *Phys. Rev. Lett.*, **59(8)**:845 (1987).
2. M. Casdagli, *Physica D*, **35**:335 (1989).
3. S. Hammel, *Phys. Lett. A*, **148**:412 (1990).
4. E. Kostelich and J.A. Yorke, *Physica D*, **41**:183 (1990).
5. R. Cawley and G. Hsu, *Phys. Rev. A*, **46(6)**:3057 (1992).
6. G. Sugihara and R. May, *Nature*, **344(N6268)**:734 (1990).
7. T. Sauer, George Mason Univ. preprint (1991).
8. T. Schreiber and P. Grassberger, *Phys. Lett. A*, **160**:411 (1991).
9. F. Takens, in Dynamical Systems and Turbulence (Springer-Verlag, Berlin, N.Y., 1981), D. Rand and L.S. Young (eds.).
10. N.H. Packard, J.P. Crutchfield, J.D. Farmer and R.S. Shaw, *Phys. Rev. Lett.*, **45(9)**:712 (1989).
11. J.P. Crutchfield and B.S. McNamara, *Complex Systems*, **1**:417-452 (1987).
12. J. Cremers and A. Huebler, *Z. Naturforschung*, **42(A)**:797 (1987).
13. J.L. Breeden and A. Hubler, *Phys. Rev. A*, **42(10)**:5817-4826 (1990).
14. J.S. Brush and J.B. Kadtke, *Proc. of the ICASSP-92*, **V**:321-325 (1992).
15. E. Baake, E., M. Baake, H.G. Bock and K.M. Briggs, *Phys. Rev. A*, **45(8)**:5524-5529 (1992).
16. M. Giona, F. Lentini, and V. Cimigalli, *Phys. Rev. A*, **45**:5524 (1992).
17. W.H. Press, S.A. Teukolsky, W.T. Vetterling, and B.P. Flannery, Numerical Recipes (Cambridge University Press, 1988).
18. J.S. Brush and J.B. Kadtke, to appear in Limits of Predictability II (Springer-Verlag, 1994-5), Yu.A. Kravtsov, J.B. Kadtke and J.S. Brush, (eds.).
19. J.S. Brush and J.B. Kadtke, *Proc. of 2nd ONR Tech. Conf. on NLD and Full Spectrum Processing* (American Inst. Physics, 1993-4).
20. Yu. A. Kravtsov, in Nonlinear Waves, Vol. 2 (Springer-Verlag, 1989), Gaponov, Rabinovich and Engelbrecht (eds.).

21. J.B. Kadtke, J.S. Brush and J. Holzfuss, to appear in *Intl. J. of Bif. and Chaos* (1993).
22. J.B. Kadtke and J.S. Brush, *Proc. of the SPIE Conf. on Sig. Proc., Sensor Fusion, and Target Recog.*, **1699**:338-349 (1992).
23. J.S. Brush, J.B. Kadtke, and Yu.A. Kravtsov, in review (preprint 1992).

ADVANCED APPLIED
SIGNAL PROCESSING METHODS

TRANSIENT DETECTION USING WAVELETS

Patricia H. Carter
Naval Surface Warfare Center
Silver Spring, MD 20901-5640

ABSTRACT

We describe the implementation of a wavelet based transient detection algorithm and compare its performance with that of traditional Fourier tecniques via a numerical simulation. The wavelet algorithm is a filter-then-detect algorithm; the filtering is accomplished in the wavelet transform domain by thresholding. The context is established by comparing the Fourier and wavelet transform in the signal processing setting. The performance of the wavelet filtering and Fourier-based filtering as a part of the detection process on two simulated detection problems is compared using ROC curves.

INTRODUCTION

Nonlinear signal processing with wavelets shares some common themes with nonlinear signal processing with tools derived from dynamical systems theory. Both present alternatives to Fourier theory. Both can be used to separate broadband signal from broadband noise - however the underlying meaning of the separation is different in the two cases. Wavelets can be used to address four basic signal processing objectives: 1) finding the signal, 2) finding the appropriate space, 3) classifying the signal and 4) making models and predicting. This work addresses the first problem only. Signal processing with wavelets is, relatively, close to Fourier theory and far from dynamics-based signal processing techniques. The time series considered here are discrete finite length sequences. Transients are signals of interest whose duration is longer than one time sample but shorter than the duration of the observation window for the time series. The transients considered here are either unknown or imperfectly known. This work is aimed toward the detection of broadband signals typical of those found in passive sonar at signal-to-noise ratios appropriate to the sonar context.

Detection of transients can be done in the time domain or in a transform domain. Time domain methods frequently have two components: filtering to enhance the signal then detection. One non-Fourier and nonlinear example of pre-detection filtering used is median subtraction. A commonly used transform for transient detection (and classification) is the Fourier transform. Detection may be accomplished directly in the Fourier domain, or the time series may be filtered in the Fourier domain, then inverse transformed back to the time domain prior to detection. In the first case, Fourier transforms of the incoming time series over successive, and overlapping, short periods of time (the length depends on the expected duration of the transient) are used; a signal may be detected, e.g., by exceedences in the power in certain frequencies. In the second case the time series over whole observation window may be Fourier transformed, and operations are performed on the transform of the time series. The altered Fourier domain data is then inverse transformed, resulting in a filtered time series. Any linear filtering-before-detection scheme fits into the second framework because linear filters have representations as convolutions, which can be implemented in the Fourier domain by multiplication of the transform of the time series and the transform of the convolution kernel.

The time domain representation of a signal is localized in time but not frequency. The frequency domain (Fourier) representation is localized in frequency but not time. A wavelet basis gives a representation of a time series which balances localization in the time against localization in frequency. (Wavelets localize in time and scale, but for this discussion we appeal to the rule of thumb that frequency $\sim$ 1/scale.) The wavelet transform, like the Fourier transform, permits a change of basis for the space of signals. In the Fourier case the basis consists of the complex exponentials, $\exp(2\pi kix)$, where k is an integer whose absolute value is less than the Nyquist frequency. In the wavelet case the basis consists of translations and dilations of a single waveform which is itself localized in time and frequency. The waveform used may be selected from a wide range of possibilities: it may be customized for the intended application or an "off-the-shelf" generic one may be used. In the numerical work described below a wavelet introduced by Daubechies[1] is used, but use of this particular wavelet is not central to the results.

Transform methods are useful in signal detection because some bases will concentrate the signal onto relatively few of the basis elements while at the same time smearing the noise around on a large number of basis elements, thus making the signal more detectable. For example, in the case of a time series consisting of a single tone plus Gaussian white noise the Fourier transform is optimal because the signal energy that was spread out over the whole observation window in the time domain is concentrated at one point the in frequency domain, while the noise energy is again spread over the whole frequency domain observation window. Short duration acoustic transients often contain a broad range of frequencies that can make detection in a noisy background difficult using Fourier based methods. However, these transients may have a compact wavelet representation, indeed their limited duration insures this to some extent.

WAVELETS

The Daubechies 4 wavelet[1] used here does not have a closed form representation but rather has a representation involving the solution of a dilation equation; numerical evaluation of it's values at the dyadic rationals has a convenient recursive formulation. Define four coefficients by $c_0=(1+\sqrt{3})/4$, $c_1=(3+\sqrt{3})/4$, $c_2=1-c_0$ and

$c_3 = 1-c_1$. The scaling function g is the solution of

$$g(x) = c_0 g(2x) + c_1 g(2x-1) + c_2 g(2x-2) + c_3 g(2x-3) \qquad (1)$$

The Daubechies wavelet W is defined in terms of the scaling function g as

$$W(x) = -c_0 g(2x-1) + c_1 g(2x) - c_2 g(2x+1) + c_3 g(2x+2) \qquad (2)$$

The collection of dyadic translations and dilations of W,

$$W_{i,k}(x) = 2^{i/2} W(2^i x - k), \qquad (3)$$

form an orthonormal basis for $L_2(\mathbb{R})$, i.e., each finite energy function has a representation of the form

$$f(x) = \sum_{i,k} c_{i,k} W(2^i x - k). \qquad (4)$$

In Figure 1a is a graph of the wavelet W, and in Figure 1b the graph of its dilation by a factor of 2. Figures 1c and 1d display their Fourier spectra (absolute amplitudes of their periodograms.)

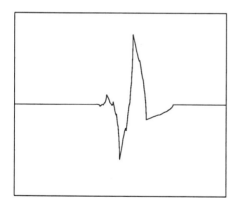

Figure 1a. W

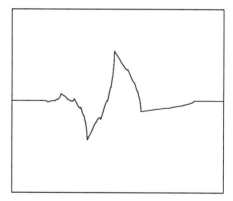

Figure 1b. $W_{-1,0}(x) = W(x/2)/\sqrt{2}$

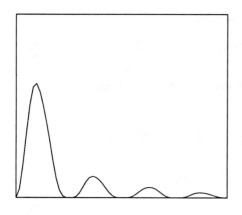 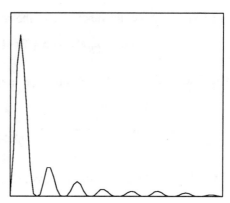

Figure 1c. amplitude spectrum of W for frequencies 1 to Nyquist/2

Figure 1d. amplitude spectrum of $W_{-1,0}$ for frequencies 1 to Nyquist/2

For the discrete transform of a finite time series of length 2^N these coefficients can be determined by an iterative procedure (pyramid algorithm) which is $O(2^N)$. The inverse transform can be implemented by a complimentary iterative procedure. These are explicitly developed in, for example, Strang [2]. Each step of the forward transform consists of a convolution followed by a decimation by two. The convolution kernels are the 4-vectors, $[c_0, c_1, c_2, c_3]$ and $[c_3, -c_2, c_1, -c_0]$. If we use discrete periodic convolution, in analogy with the usual Fast Fourier Transform implementation, then the coding is straightforward. The same caution in interpretation of results that is necessary with the FFT is required because of the wrap around effect at the endpoints.

For a time series of length 2^N, the N scales in its transform are 2^i for $i = 1, 2, ..., N$. There are $2^{(N-i)}$ coefficients at the ith scale, for a total of $2^N - 1$. The one additional coefficient is the mean. So the coefficients in equation (4) reduce to $c_{i,k}$: $k=1,...2^{(N-k)}$, $i=1,...N$ and the mean, $c_{0,0}$. One way the transform can be presented graphicly is by plotting $c_{0,0}, c_{1,1}, c_{2,1}, c_{2,2}, c_{3,1}, c_{3,2}, c_{3,3}, c_{3,4}, ..., c_{N,1}, ..., C_{N, 2N-1}$. So the right half of the plot represents the smallest scales, the adjacent fourth the second smallest scales, etc.

FILTERING IN THE WAVELET TRANSFORM DOMAIN

There are different ways hard thresholding in the transform domain can be done, resulting in either linear or nonlinear processing. The thresholding procedures discussed below are examples of hard thresholding. Other options which will not be explored here for filtering by thresholding in the wavelet transform include soft thresholding[3], nonlinear transformations of the coefficients[4-6], and representing transform by its local maxima[7-8] and then thresholding, etc.

Suppose S is a time series with wavelet transform coefficients
$$\{c_{0,0},\ c_{i,k}:\ k=1,\ldots 2^{(N-k)},\ i=1,\ldots N\}.$$
Let τ be a function of scale, i. e.,
$$\tau(i),\ i=0,\ 1,\ \ldots,\ N,$$
with $0 \leq \tau(i) \leq \infty$. The wavelet coefficients for the filtered time series $S^{\wedge}$ are given by

$$b_{i,k} = \begin{cases} c_{i,k} & \text{if } |c_{i,k}| \geq \tau(i) \\ 0 & \text{otherwise} \end{cases} \quad (5)$$

In the sequel the process of finding $S^{\wedge}$ from S will be called wavelet threshold filtering.

The first example of wavelet threshold filtering is analogous to ideal band pass filtering in the Fourier domain: in the wavelet domain, zero all the coefficients in certain scales, then inverse transform back to the time domain. If κ are the scales to be kept, then the new coefficients are given by:

$$b_{i,k} = \begin{cases} c_{i,k} & \text{if } i \in \kappa \\ 0 & \text{otherwise} \end{cases} \quad (6)$$

If there is apriori knowledge of the scales in which the signal, or the noise, is predominant, then those scales could comprise κ, or the complement of κ, respectively. This procedure is a linear filter. This corresponds with a choice of $\tau(i) = 0$ if $i \in \kappa$ and ∞ otherwise. From Figure 1b, the amplitude of the Fourier transform of the wavelet, it can be seen that this is similar to, but not, a band pass filter. A different implementation of this alternative is to use the wavelet coefficients in a single scale as the detection statistic[9]. Figure 2 shows the effect of this filtering: Figure 2a is the graph of a signal plus noise time series with 1024 points. The signal is the broadband transient discussed in the simulation section, the noise is Gaussian white noise and the signal to noise ratio (as defined in the simulation section) is 3 dB. Figure 2b shows the squares of its wavelet transform coefficients, arranged in the order from largest to smallest scales as explained above. Figure 2c shows the new wavelet coefficients retaining only scales 4 through 7. Figure 2d shows the resulting filtered time series. In the sequel this special case of wavelet threshold filtering will be called linear wavelet filtering.

238 Transient Detection Using Wavelets

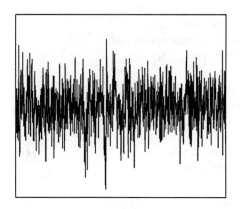

Figure 2a. original time series

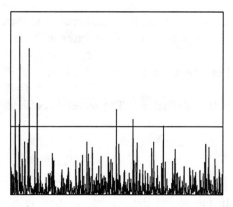

Figure 2b. wavelet coefficients squared with noise threshold

Figure 2c. new wavelet coefficients

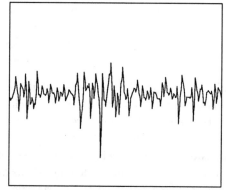

Figure 2d. filtered time series

The second example of wavelet threshold filtering is that of applying a single threshold τ to all the coefficients of the wavelet transform of a time series. So $\tau(i) = \tau$ for all i in equation (5). The new wavelet coefficients are

$$b_{i,k} = \begin{cases} c_{i,k} & \text{if } |c_{i,k}| \geq \tau \\ 0 & \text{otherwise} \end{cases} \qquad (7)$$

Inverse transforming these coefficients to obtain a new time series produces a nonlinear filtering of the original time series. DeVore and Lucier[10] show this can be used to produce a near-optimal separation of signal and noise, in the case in which the noise is Gaussian. For a numerical implementation, a determination of a reasonable threshold must be made. For the example shown in Figure 3 the threshold $\tau=2.8\sigma$, where σ is the standard deviation of the noise. The threshold is indicated by the horozontal line at height τ^2 in Figure 2b. Figure 3 shows the effect of this filtering on the signal plus noise shown in Figure 2a: Figure 3a shows the squares of the wavelet transform coefficients after the threshold is applied, and 3b shows the resulting filtered time series. This will be referred to as unbiased wavelet filtering.

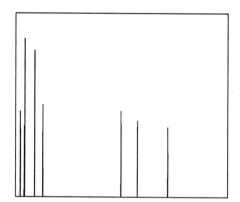 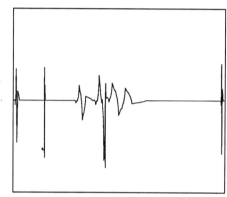

Figure 3a. new wavelet coefficients   Figure 3b. filtered time series

These two methods can be used together. Figure 4 shows the combination of keeping the same scales as Figure 2, then thresholding them as done to produce Figure 4: Figure 4a shows the squares of the wavelet coefficient after both operations and 4b shows the resulting filtered time series. In the sequel this will be referred to as mixed wavelet filtering.

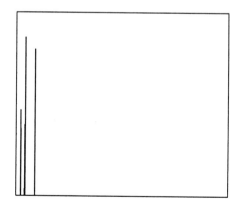 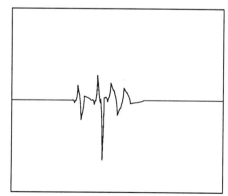

Figure 4a. new wavelet coefficients   Figure 4b. filtered time series

Suppose that signal free examples of the noise are available. Consider, for some constant m,

$$\tau(i) = m\,\sigma(i) \quad for\ i = 1,...,N \qquad (8)$$

where $\sigma(i)$ is the standard deviation of the wavelet transform of the noise at the ith scale. This threshold uses no information about the signal. Schiff[11] uses essentially this thresholding, calculating the σ's from a randomized version of his signal plus noise rather than a signal-free noise samples. In the case of Gaussian noise the numbers $\sigma(i)$ are all equal to the standard deviation of the noise, so this reduces to the single number thresholding in equation 7.

The performance of these three wavelet transform filtering schemes: linear wavelet filtering, unbiased wavelet threshold filtering and mixed wavelet filtering, will be compared with standard linear filtering techniques: 1) the matched filter (replicate correlator), 2) a power spectrum-matched filter, and 3) a band pass filter. These all require some apriori knowledge of the signal to be detected: the matched filter requires exact knowledge of each time sample of the signal. The power spectrum-matched filter requires knowledge of the power spectrum of the signal (essentially the amplitude but not the phases of the Fourier transform of the signal). The band energy detector requires a band of frequencies which contains most of the signal power. In contrast, the linear wavelet filtering uses knowledge about the mean power in each scale of the wavelet transform of the signal. The unbiased wavelet filtering uses the standard deviation of the noise, and the mixed wavelet filtering uses both. We also assume (in not using the largest two wavelet scales) that the signal's duration is less than about 1/4 of the observation window. The six filtering methods have parallel implementions: for each the original signal is transformed, filtered in the transform domain, and then transformed back to the time domain prior to detection. Note that the replicate correlator is optimal among *linear* filters in this context.

THE SIMULATION

This is a Monte Carlo simulation of the binary detection problem, with unknown onset of signal. The signals used are: 1) a broadband synthetic transient, and 2) a synthetic bang. The graphs of the non-zero part of each of the signals in the time domain are in Figures 5a and 5c. Their amplitude spectra are shown in Figures 5b and 5d.

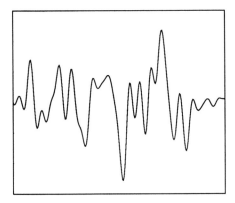

Figure 5a. broadband signal

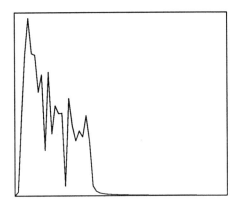

Figure 5b. amplitude spectrum broadband signal, frequencies 1 to Nyquist/2

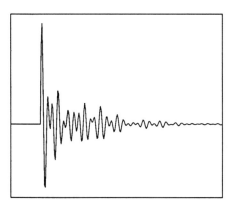

Figure 5c. synthetic bang

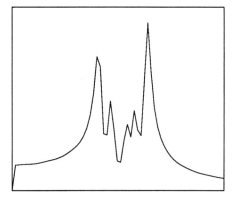

Figure 5d. amplitude spectrum synthetic bang, frequencies 1 to Nyquist/2

The noise is white Gaussian, with 0 mean and variance σ^2. Each signal is added in turn to different realizations of the noise, where σ^2 chosen to achieve the stated signal-to-noise ratio (SNR). The SNR is defined to be

$$SNR = 10\log_{10}\frac{\|sig(t)\|_2 / M}{\|nfil(t)\|_2 / 2^N} \tag{9}$$

where M is the duration in samples of the signal, 2^N is the size of the observation window, sig is the signal, nfil is the noise, filtered to the band containing 99% of the signal power, and the l_2 norm is $|f(t)|_2 = \sum f(t)^2$. This SNR is a measure of the ratio of the signal energy per sample over the in-band noise energy per sample. This is a fairly strict definition of SNR, taking into account both the time duration and frequency band of the signal.

The effectiveness of a detection algorithm depends the probability of false alarm as well as the probability of detection at a given threshold. The two situations of interest are:

signal declared   -   signal present
signal declared   -   no signal present

For any detection algorithm increasing the probability of detection necessarily increases the probability of false alarm. The relationship between the two is given by the receiver operating characteristic (ROC) curve. ROC curves are one way to quantify and compare the performance of detection algorithms on synthetic and real data For each τ a point (pfa(τ), pd(τ)) is plotted on the ROC curve where: pfa(τ) is the probability of false alarm when the threshold is set at τ, and pd(τ) is the probability of detection when the threshold is set at the same τ.

For each of the methods the detection statistic used is the square of the filtered signal, smoothed over a window of size 32 and subsampled by a factor of 8. For each realization of the noise, given one threshold value, the detection statistic is compared with that value sample by sample. If the statistic lies on or above the threshold then a "signal" is declared at that sample, if below then "no signal" is declared at that sample. This process is repeated over many realizations of the noise to determine the probabilities of detection and false alarm. The resulting ROC curves are displayed here in log-log plots. Comparing two ROC curves the one which is closer to the top left corner is the one with better performance.

For this study the simulated signals were 256 samples long, and the observation window 4096 samples. There were 1000 realizations of Gaussian white noise used. A set of 128 thresholds for the detection statistic were used to determine points on the ROC curves..

For the linear wavelet filtering simulation, information about the signal is used to determine κ. The mean total energy in each scale for the broadband signal and for the synthetic bang are shown in Figure 7. The mean is over the possible translations of the 256 sample signal in the window of observation. The scales in $\kappa_b = \{6, 7, 8\}$ contain 80% of the broadband signal's energy and the scales in $\kappa_s = \{9, 10\}$ contain 83% of the energy of the synthetic bang.

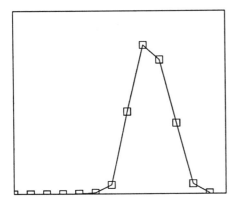 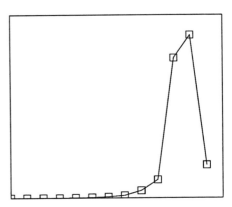

Figure 6a. mean total energy in each scale for broadband signal

Figure 6b. mean total energy in each scale for synthetic bang

For the unbiased wavelet filtering the threshold is set at $\tau=2.5\sigma$, where σ is the standard deviation of the noise at the given SNR. For the broadband signal the SNR is 0 dB and for the synthetic bang the SNR is -8. For the mixed filtering, in the broadband case the scales κ_b are kept and thresholded at $\tau=2.5\sigma$, and in the synthetic bang case, the scales κ_s are kept and thresholded at $\tau=2.5\sigma$. The ROC curves resulting from the two simulations are shown in Figures 7a and 7b.

For the broadband signal at 0 dB the methods in increasing order of effectiveness are the unbiased wavelet, the band pass, the linear wavelet, the matched power spectrum, the mixed wavelet and the replicate correlator. The choices of threshold and number of realizations used did not seem to allow exploration of performance when the probability of false alarm is less than .001 or where the probability of detection is close to 1 for the mixed wavelet case.

For the synthetic bang at -8 dB, the wavelet unbiased filtering, the wavelet linear filtering and the band pass filtering are performing similarly, with the band pass maybe slightly worse. The replicate correlator and the power spectrum matched filters are performing similarly. The mixed wavelet is just below them, for probabilites of false alarm greater than .001. However, again, with the thresholds chosen and the number of realizations used the performance of the mixed wavelet for lower false alarm rates is not explored.

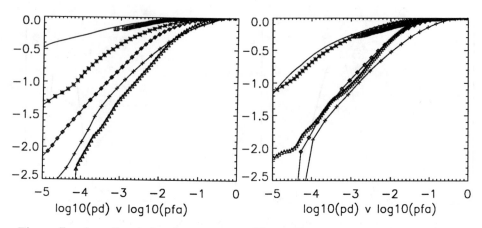

Figure 7a. broadband signal   Figure 7b. synthetic bang

Figure 7. ROC curves

symbol legend:

band pass +   matched power spectrum *   replicate correlator −
linear wavelet ◇   unbiased wavelet ▵   mixed wavelet ☐

CONCLUSIONS

Filtering using thresholding in the wavelet transform domain is a promising new technique which has a powerful theory behind it. The question addressed here was how well does it work in practice. The implementation using some information about the signal and some information about the noise produced surprisingly good results in the simulation. A systematic exploration, over various noise and signal types, as well as over various thresholding protocols, of its performance in detection algorithms has been started. This systematic study needs to be finished before any firm conclusions can be reached on its merit relative to established Fourier techniques.

ACKNOWLEDGEMENTS

This work was supported by ONR Code 23, Tommy Goldsberry, under the project Evaluating Nonlinear Dynamics.

REFERENCES

1. I. Daubechies, Ten Lectures on Wavelets, SIAM, Philadelphia, PA (1992)

2. G. Strang, SIAM Review **31** p. 614-627 (1989)

3. D. L. Donoho, De-Noising via Soft Thresholding, TR 409 preprint (November 1992)

4. D. L. Donoho, I. M. Johnstone, Ideal Spatial Adaptation via Wavelet Shrinkage, TR 400 preprint (July 1992)

5. D. L Donoho, I. M. Johnstone, Minimax Estimation via Wavelet Shrinkage, TR 402 preprint (July 1992)

6. D. L. Donoho, Unconditional Bases are Optimal Bases for Data Compression and for Statistical Estimation, TR 410 preprint (November 1992)

7. S. G. Mallat, IEEE Trans. Pattern Analysis and Machine Intell., **11** (1989)

8. S. Zhong, S. G. Mallat, Compact image representation from multiscale edges, 3rd International Conference on Computer Vision, Japan (1990)

9. R. Carmona, Wavelet detection of transients in noisy time series, preprint (1992)

10. R. A. Devore, B. J. Lucier, 1992 IEEE Military Communications Conference, p. 48.3.1-48.3.7, IEEE Communications Society (1992)

11. S. J. Schiff, Opt. Eng. **34** p. 2492-2495 (1992)

DYNAMICAL SYSTEMS WITH CYCLOSTATIONARY ORBITS

Harry L. Hurd and Carl H. Jones
Harry L. Hurd Associates, 2301 Stonehenge Dr., Suite 104
Raleigh, NC 27615

Abstract

In an attempt to better understand physical mechanisms that generate cyclostationary processes we were lead to consider orbits generated by periodically perturbing a parameter of a dynamical system. A physical example is the effect of the rotation of the earth on meteorological processes. Using statistical tests for the presence of cyclostationarity, we show that orbits of the periodically perturbed logistic map are consistent with the presence of cyclostationarity. Further, the orbit of an unperturbed system also indicates the presence of cyclostationarity. In contrast to the power spectrum, the spectral correlation methods used for determining the presence of cyclostationarity utilize both amplitude and phase of the sample Fourier transform.

INTRODUCTION

Many models of physical processes (time functions) are assumed to possess the quality of stationarity; measurements made on the statistical properties of the processes are assumed to be invariant with respect to arbitrary time shifts. For example, the probability distributions of strictly stationary stochastic processes are invariant with respect to time shifts. And in the case of second order processes, the correlation

$$R(s+\tau, t+\tau) = E\{X(s+\tau)\overline{X(t+\tau)}\} \qquad (1)$$

of a wide-sense stationary process is invariant with respect to the shift τ for every s, t. This leads easily to the conclusion that $R(s,t)$ depends only on the difference $s-t$.

There is another class of processes, called *cyclostationary*[1] , that arise through the assumption that the statistical properties of the processes are invariant with respect to shifts of size T, but for no smaller values. Then T is called the period of the cyclostationary process. To be precise, a stochastic process $\{X(t), t \in \mathbf{R}\}$ is called *strictly cyclostationary* if

$$Pr[X(t_1) \in F_1, \ldots, X(t_n) \in F_n] = Pr[X(t_1+T) \in F_1, \ldots, X(t_n+T) \in F_n] \qquad (2)$$

for arbitrary n, times $\{t_j, j = 1, \ldots n\}$ and Borel sets $\{F_j, j = 1, \ldots n\}$. Similarly, a second order process $\{X(t), t \in \mathbf{R}\}$ is called *cyclostationary* or *periodically correlated* if

$$R(s,t) = R(s+T, t+T) \qquad (3)$$

for every s, t. These processes occur, for example, when stationary processes are subjected to periodic modulations, either in the amplitude or time-scale. Some basic references are [1, 2, 3]. For a complete collection of references see [4].

---
[1]These processes have also been called periodically stationary, periodically nonstationary, T-stationary and in the second order case, periodically correlated.

In this paper we shall review some of the spectral properties of second order cyclostationary *sequences* and describe tests for the presence of cyclostationarity based on the spectral theory. We then show that these tests indicate, with considerable significance, the presence of cyclostationarity in the orbits of a periodically perturbed logistic map and in the orbits of an unperturbed two-dimensional map arising from an age structured population model[5].

Although we often use the synonymous term *periodically correlated* to describe second order cyclostationary sequences, we defer here to the work of Gardner[2] in which a property of cyclostationarity is assigned to one specific deterministic function. This may be seen as an extension of Wiener's generalized harmonic analysis where a property of stationarity is assigned to a single function. As in generalized harmonic analysis, the definition of cyclostationarity rests on the existence and properties of certain time averages, and under reasonable assumptions of ergodicity[2], these time averages will all exist for almost all sample paths of a cyclostationary *random* sequence. But in the other direction, when we are given a single deterministic time series the problem of making probabilistic statements becomes more difficult because we are faced with determining a measure space based on a single time series. In this direction, Gardner[2] has introduced cyclic "fraction of time" probabilistic models, and also shown that measures of consistency of certain estimators may be expressed in terms of time averages. Related problems motivated by Gardner's work are treated in [7]. Here we shall take the view that we are simply reporting on the application of a statistical test for cyclostationarity to a time series generated by a deterministic (non-random) system.

BACKGROUND

As one may see from the preceding definitions, cyclostationary sequences are generally not stationary and contain the stationary sequences as a subset. The mathematical basis for the tests for cyclostationarity described herein is the concept of *harmonizable* random sequences and processes. In this paper we shall concentrate on sequences, but similar statements can be made about continuous time processes. Harmonizable processes were introduced by Loève [8] through a generalization of the fact that every wide-sense stationary process has an integral spectral representation of the form

$$X_n = \int_0^{2\pi} \exp(i\lambda n) \, dZ(\lambda) \qquad (4)$$

where the frequency-indexed random process $Z(\lambda)$ has orthogonal or uncorrelated increments. In contrast, for a nonstationary *harmonizable* process the increments need not be orthogonal. The exact manner in which the increments of $Z(\lambda)$ are not orthogonal gives some information about the nature of the nonstationarity of X_n. For example, if X_n is periodically correlated (cyclostationary), the increments $dZ(\lambda_1)$ and $dZ(\lambda_2)$ have non-zero correlation only when $\lambda_2 = \lambda_1 - 2\pi k/T$ for $\lambda_1 \in [0, 2\pi)$. This fact may also be described in terms of the support set of the spectral correlation measure given formally

---

[2] The first discussion of this problem appears in Boyles and Gardner [6]; for an update see [4]

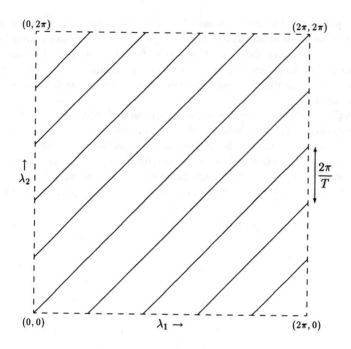

Figure 1: Support of r_Z for a Periodically Correlated Sequence.

by $r_Z(d\lambda_1, d\lambda_2) = E\{dZ(\lambda_1)\overline{dZ(\lambda_2)}\}$, where $r_Z(d\lambda_1, d\lambda_2)$ is related to the correlation function of X_n by

$$E\{X_m \overline{X_n}\} = R(m,n) = \int_0^{2\pi} \int_0^{2\pi} \exp(i\lambda_1 m - i\lambda_2 n)\, r_Z(d\lambda_1, d\lambda_2). \qquad (5)$$

Then it is known[9] that X_n is periodically correlated if and only if the support of $r_Z(d\lambda_1, d\lambda_2)$ is contained in the intersection of the $2T-1$ diagonal lines $\{(\lambda_1, \lambda_2) : \lambda_2 = \lambda_1 - 2\pi k/T, k = -(T-1), \ldots, T-1\}$ with the square $[0, 2\pi) \times [0, 2\pi)$. This support set is illustrated in Figure 1. For a review of the spectral theory of cyclostationary processes see Yaglom[10]; for a more recent survey that includes the almost periodic case, see Dehay and Hurd[11].

Since the (spectral) correlations in the increments of Z that give rise to Figure 1 is characteristic of cyclostationary sequences, then it follows that a test for the presence of cyclostationarity may be phrased in terms of a test for these specific correlations in the sample spectrum. Empirically, the sample Fourier transform may be interpreted as an estimate of the frequency increments of a harmonizable process.

This idea originated with N.R. Goodman[12] who proposed several statistical tests for determining the presence of nonstationarity (in the context of harmonizable processes) by testing for the presence of correlation in estimates of the increments. Recently it has been noticed that one of Goodman's tests, based on the sample coherence statistic, (actually diagonal spectral coherence) provides a natural test for cyclostationarity[13].

We now review the spectral coherence computation and it's use for determining the presence of cyclostationarity.

The first step in the computation of diagonal spectral coherence is the calculation of the finite length sample Fourier transform of the sequence X_n. That is,

$$Z_k = \sum_{n=0}^{N-1} X_n \exp(-i2\pi kn/N) \qquad (6)$$

where we interpret the random variables Z_k as estimates of $dZ(\lambda)$ at the finite set of frequencies given by $\lambda_k = 2\pi k/N$. The spectral coherence computation determines the normalized correlation, or coherence, between two bands $Z_p, Z_{p+1}, \ldots Z_{p+M-1}$ and $Z_q, Z_{q+1}, \ldots Z_{q+M-1}$ of frequency-indexed variates where each band contains M variates indexed consecutively. Precisely, we compute

$$|\gamma(p,q,M)|^2 = \frac{|\sum_{m=0}^{M-1} Z_{p+m}\overline{Z_{q+m}}|^2}{\sum_{m=0}^{M-1} |Z_{p+m}|^2 \sum_{m=0}^{M-1} |Z_{q+m}|^2} \qquad (7)$$

for (p,q) in a square array and plot the values exceeding a threshold according to a greyscale encoding. A useful threshold is determined by the null distribution of $|\gamma(p,q,M)|^2$ under the assumption that the Z_k are i.i.d. complex Gaussian variates, and that any Z_{p+m} in the numerator of (7) does not appear as a Z_{q+m}. Then $Pr[|\gamma|^2 > |\gamma_0|^2] = (1 - |\gamma_0|^2)^{M-1}$ (see [12]).

The spectral coherence computation may also be interpreted as a uniform smoothing of the two-dimensional periodogram

$$g(N,j,k) = \frac{1}{2\pi N} Z_j \overline{Z_k} \qquad (8)$$

along a diagonal line having unit slope; the smoothed two-dimensional periodogram is then normalized by the product of the smoothed diagonal terms. For this reason this particular coherence calculation is called *diagonal spectral coherence*. Since the support of the spectral correlation measure r_Z for cyclostationary sequences consists of straight lines of unity slope, the diagonal spectral coherence computation gives a test for the presence of cyclostationarity[13].

A computer program for the computation of diagonal spectral coherence, along with other related programs for the time series analysis of cyclostationary signals has been integrated into one master program called DSCOH.[3] Figure 2a presents an example of a spectral coherence image produced from a time series that was comprised of simulated white noise; the locations of high coherence are randomly distributed throughout the image. In contrast, Figure 2b is a spectral coherence image produced by a simulated cyclostationary sequence generated by periodic modulation of white noise. In both cases the plotting threshold corresponds to .05 significance level[4] at each point (p,q). It may be easily determined from (7) that on the main diagonal the spectral coherence is unity: $|\gamma(p,p,M)|^2 = 1$. The value of the period T may be inferred from the SC images by determining the least value of $p-q$ for which there is a clear indication of correlation.

---

[3] Additional information about the DSCOH program may be obtained from the authors.
[4] Also called the probability of false alarm, and also the probability of Type 1 error.

250 Dynamical Systems with Cyclostationary Orbits

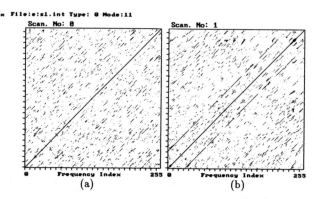

Figure 2: (a) - Diagonal Spectral Coherence Image for White Gaussian Noise. (b) - Diagonal Spectral Coherence Image for Simulated Cyclostationary Sequence.

The image may be interpreted differently for small and large values of M. For small M, a threshold exceedance (a visible dot on the image) says that the two small bands of frequency components, corresponding to the coordinates (p,q) of the dot, have significant coherence. The noise only case of Figure 2a shows many such dots (roughly 5% of all the coordinate pairs have visible dots) but none show persistence along a particular diagonal line as they do in Figure 2b. So for small M, we evidently require more than the occurrence of isolated threshold exceedances to be convinced of the presence of cyclostationarity. The requirement of persistence along fixed diagonal lines is one way to combine many local tests of $|\gamma|^2$ into a statement that a support line exists at some value of d, and hence cyclostationarity is present.[5] The persistence requirement may be viewed as an averaging process that reduces the probability of falsely declaring cyclostationarity based on the local $|\gamma|^2$ tests.

Another method of reducing the probability of falsely declaring cyclostationarity (in comparison to the local $|\gamma|^2$ tests) is to maintain the $|\gamma|^2$ threshold of the local test, increase M and then declare cyclostationarity whenever the threshold is exceeded at any point (p,q). The expression $Pr[|\gamma|^2 > |\gamma_0|^2] = (1-|\gamma_0|^2)^{M-1}$ shows how the probability of type 1 error decreases as M increases. Thus for larger values of M, the occurrence of any dot may be used as a local test for cyclostationarity.

In the limit we may make M as large as permitted (i.e., $M = N$) so that only one value of spectral coherence is determined for each separation d from the main diagonal. Thus we compute $|\gamma(0,d,N)|^2$ where N is the dimension of the Fourier transform vector, and frequencies are taken modulo 2π as appropriate. We call $|\gamma(0,d,N)|^2$ the *coherent* statistic because all the Z_k are used in one coherence calculation for each d. The numbers $|\gamma(0,d,N)|^2$ are plotted as a function of the parameter d, which corresponds to the distance from the main diagonal in the bi-frequency coherence plots, and is a difference frequency. It is shown by Bloomfield and Hurd[14] that for $M = N$ the numerator of (7) is given by

$$\sum_{m=0}^{N-1} Z_{p+m} \overline{Z_{q+m}} = N \sum_{n=0}^{N-1} |X_n|^2 \, exp[i2\pi(q-p)n/N]; \qquad (9)$$

---

[5] The issue of testing for the presence of almost cyclostationarity will not be treated here.

thus it may be seen that $|\gamma(p,q,M)|^2$, as a function of d, is proportional to the magnitude of the normalized Fourier transform of $|X_n|^2$. This considerably reduces the number of computations required to produce the coherent statistic.

Choosing $M = N$ can sometimes be a detriment because it is possible for the phase of the spectral correlation measure r_Z in (5) to change along it's support line. Thus it is possible that the quantity (7) may average to zero; a simple example is discussed in Section 5 of [13] and a related example is given below. This argues that smaller values of M also have their purpose. To obtain a more powerful test for small values of M that still requires persistence along diagonal lines, we *incoherently* average all the values of $|\gamma(p,q,M)|^2$ from a coherence plot for a fixed value of $d = p - q$. To be precise, the *incoherent* statistic is the average

$$\delta(d,M) = \frac{1}{N_i - 1 - d} \sum_{p=0}^{N_i-1-d} |\gamma(p,p+d,M)|^2, \qquad (10)$$

where N_i is the dimension of the displayed image. The quantity $\delta(d,M)$ is plotted as a function of the difference frequency d. In essence, this statistic was utilized by Bloomfield, Hurd and Lund[15] for determining whether residuals of time series after model-fitting are cyclostationary.

We have now described three methods for determining the presence of cyclostationarity: the "raw" spectral coherence image, the coherent statistic and the incoherent statistic. All three use both amplitude and phase of the complex sample Fourier transform. The *coherent* statistic results from one long coherence computation using all the Z_k (i.e., $M = N$) for each value of d. In contrast, the *incoherent* statistic is the (incoherent) average of $|\gamma|^2$ values, computed with $M < N$, found along a fixed diagonal line.

Figure 3 is a combined display produced by DSCOH containing the time series, the spectral coherence image, the periodogram and plots of the incoherent and coherent

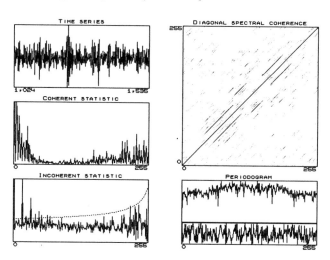

Figure 3: Combined Display with Coherent and Incoherent Statistics, Frequency Modulated Shaped Noise, Pr[Incoherent Statistic > Threshold]=.001, N=512, M=16.

Table I: Values of ν_M Determined by Simulation.

M	ν_M
2	.1858
4	.1758
6	.1577
8	.1277
10	.1196
12	.0978
16	.0744
24	.0533

statistics. The signal was a frequency modulated noise (not white) and produces an example in which the spectral correlation measure r_Z integrates to zero along the diagonal lines corresponding to the modulation frequency. The spectral coherence image shows support lines at the spacing corresponding to the modulation frequency, and corresponding peaks are seen in the plot of the incoherent statistic. They are not seen in the plot of the coherent statistic because of the rapid fluctuations in r_Z along the support line.

The dotted line that overlays the plot of the incoherent statistic $\delta(d, M)$ is a threshold corresponding to a significance level of .001 for a Gaussian null distribution with mean $1/M$ and variance $\sigma(M, N) = \nu_M/N$. The normality is based on a central limit argument. Table I gives values for ν_M determined via simulation[16]. Additional recent results on testing for cyclostationarity are given in [17].

DYNAMICAL SYSTEMS WITH CYCLOSTATIONARY ORBITS

In order to understand how physical systems generate signals that are cyclostationary (or that have diagonal spectral coherence) we have conducted some simple experiments with periodically perturbed dynamical systems, primarily with the periodically perturbed logistic map. To be precise, given a family $g_\alpha : \mathbf{R} \to \mathbf{R}$ of maps, and a finite collection of parameters $\{\alpha_1, \alpha_2, \ldots, \alpha_{T-1}\}$, a periodically perturbed map (or family of maps) is the system defined by

$$\begin{aligned}
x_1 &= g_{\alpha_0}(x) \\
x_2 &= g_{\alpha_1} \circ g_{\alpha_0}(x) \\
&\vdots \\
x_T &= g_{\alpha_{T-1}} \circ \ldots \circ g_{\alpha_1} \circ g_{\alpha_0}(x) \\
&\vdots \\
x_n &= g_{\alpha_r} \circ g_{\alpha_{r-1}} \circ \ldots \circ g_{\alpha_0} \circ g_{\alpha_{T-1}} \circ \ldots \circ g_{\alpha_0}(x)
\end{aligned} \quad (11)$$

where $r = (n - 1) \bmod T$.

So a periodically perturbed map may be seen as an orbit of a cycle of compositions. Since the dynamics of any g_α depend on the value of α, it is of particular interest

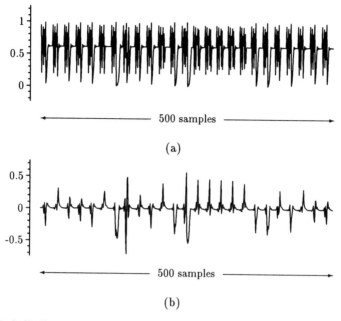

Figure 4: Periodically Perturbed Logistic Map, $T = 20, \alpha_0, \ldots \alpha_9 = 4$, $\alpha_{10}, \ldots \alpha_{19} = 2.6$, (a) The original orbit x_n. (b) The de-meaned orbit $x_n - \mu_n$.

to examine the nature of the orbit x_n as a function of the itinerary of the parameters $\{\alpha_1, \alpha_2, \ldots, \alpha_{T-1}\}$. This general question along with additional facts about the dynamics of these maps will be reported on subsequently[18].

The primary point to be demonstrated here is that the orbits of these systems possess properties (spectral coherence and others) that are consistent with those of cyclostationary sequences. To illustrate using the logistic map $g_\alpha(x) = \alpha x(1 - x)$, we have chosen a period $T = 20$ and set $\alpha_0 = \ldots = \alpha_9 = 4$ and $\alpha_{10} = \ldots = \alpha_{19} = 2.6$. So for half the period α is set to a value corresponding to chaotic behavior (in g_α) and for the other half of the period it is set to a value corresponding to stable behavior ($g_{2.6}$ has a single attracting fixed point at $x = 8/13$). The orbit in Figure 4a resulting from a computer simulation of this system appears to exhibit a periodic regularity of the sort associated with cyclostationarity although it is also clear that the orbit (in this time-frame) is not periodic with period $T = 20$. However, since the perturbation has period $T = 20$, it is considered likely that the orbit will, for sufficiently large n, have an additive periodic component with period $T = 20$. The sequence plotted in Figure 4b is $x_n - \mu_n$ where

$$\mu_n = \frac{1}{N_p} \sum_{j=1}^{N_p} x_{n+jT} \tag{12}$$

is the empirical periodic mean of period T based on a sequence of length $N_p T = 32768$.

The dynamical behavior of x_n, μ_n and $x_n - \mu_n$ within a typical interval of length $T = 20$ may be seen in Figure 5. The orbit x_n behaves rather chaotically (qualitatively)

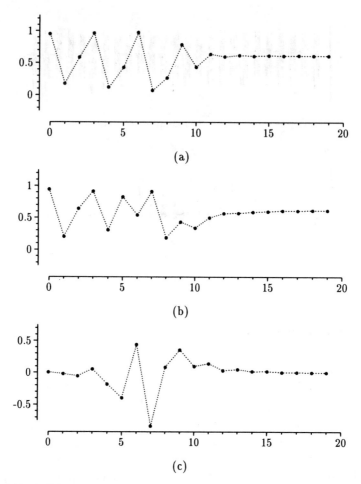

Figure 5: A Typical Cycle from the Periodically Perturbed Logistic Map, $T = 20, \alpha_0, \ldots \alpha_9 = 4$, $\alpha_{10}, \ldots \alpha_{19} = 2.6$, (a) The original orbit x_n. (b) The empirical mean μ_n. (c) The de-meaned orbit $x_n - \mu_n$. Indices are shown mod 20.

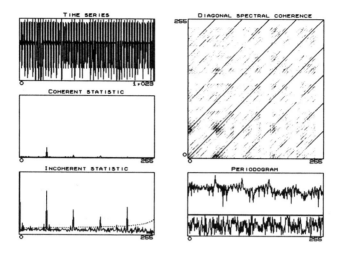

Figure 6: Periodically Perturbed Logistic Map, $T = 20, \alpha_0, \ldots, \alpha_9 = 4$, $\alpha_{10}, \ldots, \alpha_{19} = 2.6$. Combined Display of x_n with Pr[Incoherent Statistic > Threshold]=.001, $N = 1024, M = 16$.

in the first half of each cycle, and then appears to be rapidly converging in the second half. The small values of $x_n - \mu_n$ for $0 \le n \le 4$ shows that the orbit is rather stable in this interval from one period to the next, even though $\alpha = 4$ in this region. The reason for the stability is that $x_{20(k+1)}$ is always determined by $g_4(x_{19+20k})$ where x_{19+20k} is always very close to 8/13, the fixed point for $g_{2.6}$. This stability diminishes for larger n because of the sensitive dependence of g_4.

Figure 6 is a combined display of a segment of the original orbit x_n. The SC image and incoherent statistic are clearly consistent with the presence of cyclostationarity. However time series with strong additive periodic components, as seen in the periodogram of Figure 6, always produce high values of spectral coherence. This is the reason for removal of the periodic mean.

In contrast, these harmonics are absent from the periodogram of the de-meaned series, $x_n - \mu_n$ shown in Figure 7, and still the SC image and the plot of the incoherent statistic are consistent with the presence of cyclostationarity.

This evidence of cyclostationarity in periodically perturbed systems immediately brings a question concerning the existence of cyclostationarity in unperturbed systems. A preliminary investigation of a two-dimensional map given by Cazelles and Ferriere[5],

$$\begin{aligned} x_2 &= \rho y_1 \exp(-0.01(y_1 + x_1)) \\ y_2 &= 0.2x_1 \exp(-0.01(y_1 + x_1)) + 0.8y_1 \exp(-0.05(y_1 + 0.5x_1)) \end{aligned} \quad (13)$$

also produces orbits having spectral coherence that is consistent with cyclostationarity. Figure 8a is an orbit plot of the y component of the map (13) for $\rho = 800$. Qualitatively, the y-component of the orbit appears to have the required periodic regularity, although in this case we do not precisely know the period (or even if it is integral) because the system was not being perturbed at a fixed known period. Thus the simple process of

256 Dynamical Systems with Cyclostationary Orbits

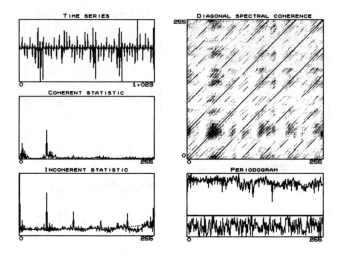

Figure 7: Periodically Perturbed Logistic Map, $T = 20, \alpha_0, \ldots, \alpha_9 = 4$, $\alpha_{10}, \ldots, \alpha_{19} = 2.6$. Combined Display of $x_n - \mu_n$ with Pr[Incoherent Statistic > Threshold]=.001, $N = 1024, M = 16$.

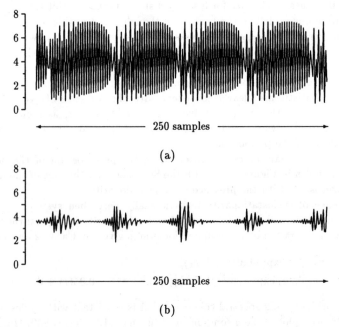

Figure 8: Two Dimensional Map, $\rho = 800$, (a) The original orbit y_n. (b) The notch-filtered orbit.

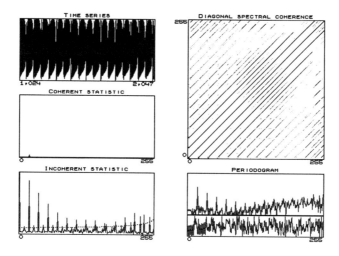

Figure 9: Combined display of the original orbit y_n with Pr[Incoherent Statistic > Threshold]=.001, $N = 1024$, $M = 16$.

removing the empirical periodic mean was not effective. In this case we first use a very long long FFT to estimate the period and then use a sequence of notch filters to remove each of the harmonics. Figure 8b is an orbit plot with the periodic mean removed in this way. Figures 9 and 10 are combined DSCOH plots that show significant levels of spectral coherence consistent with cyclostationarity.

CONCLUSIONS

This paper has shown that the orbits of periodically perturbed dynamical systems exhibit spectral properties that are consistent with cyclostationarity. To be more precise, the null hypothesis (not cyclostationary) of a statistical test based on spectral coherence is rejected at a level of significance smaller than .001. This same test also shows that the orbit of an unperturbed two dimensional map is cyclostationary. In both of these cases the strongest periodic components are first removed so they do not influence the outcome of the tests.

These observations raise some questions. What are the general properties of periodically perturbed systems that produce cyclostationary orbits? What kinds of unperturbed systems generate cyclostationary orbits and under what circumstance can they be expected to do so? To what extent can the spectral correlation methods used in the analysis of cyclostationarity contribute to the classification of orbits?

258 Dynamical Systems with Cyclostationary Orbits

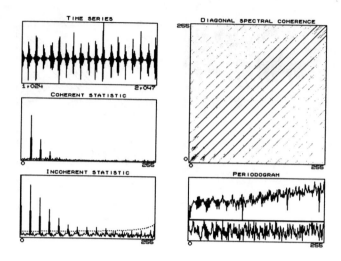

Figure 10: Combined display of the notch-filtered orbit with Pr[Incoherent Statistic > Threshold]=.001, $N = 1024, M = 16$.

ACKNOWLEDGMENTS

This work was supported by the Office of Naval Research code 1111 under contract N00014-92-C-0057. We also acknowledge the help of Fred Heller in the programming of the periodically perturbed logistic map, and Robert Lund for the data appearing in Table I.

REFERENCES

[1] W.A. Gardner, Introduction to Random Processes with Application to Signals and Systems (McMillan, New York, 1985).

[2] W.A. Gardner, Statistical Spectral Analysis: A Nonprobabilistic Theory (Prentice Hall, Englewood Cliffs, NJ, 1987).

[3] H.L. Hurd, IEEE Trans. Inform. Theory 35, 350-359 (1989).

[4] W.A. Gardner, ed. Proceedings of the Workshop on Cyclostationarity in Communications and Signal Processing (IEEE Press, 1993).

[5] Cazelles and Ferriere, Nature, 355 (Jan 2, 1992).

[6] R.A. Boyles and W.A. Gardner, IEEE Trans. Inform. Theory 29(1), 105-114 (1983).

[7] H.L. Hurd and T. Koski, The Wold Isomorphism and Cyclostationary Processes (In preparation).

[8] M. Loève, Probability Theory (Van Nostrand, N.Y., 1965).

[9] E.G. Gladyshev, Sov. Math. 2, 385-388 (1961).

[10] A.M. Yaglom, Correlation Theory of Stationary and Related Random Functions (Springer-Verlag, N.Y., 1987).

[11] D. Dehay and H.L. Hurd, Representation and Estimation for Periodically and Almost Periodically Correlated Random Processes, (to appear in Cyclostationarity in Communications and Signal Processing, W.A. Gardner, ed., IEEE Press, 1993).

[12] N.R. Goodman, Statistical Tests for Nonstationarity within the Framework of Harmonizable Processes, (Rocketdyne Research Report AD619270, 1965).

[13] H.L. Hurd and N.L. Gerr, J. Time Ser. Anal. 12(4), 337-350 (1991).

[14] P. Bloomfield and H.L. Hurd, Fourier Relationships and Nonparametric Spectral Estimation for Periodically Correlated Sequences (In preparation).

[15] P. Bloomfield, H.L. Hurd and R. Lund, Periodic Correlation in Stratospheric Ozone Time Series (To appear in J. Time Ser. Anal.).

[16] P. Bloomfield, H.L. Hurd, R. Lund and R. Smith, Two Problems of Inference for Periodically Correlated Sequences (In preparation).

[17] A.V. Dandawaté and G.B. Giannakis, Statistical Tests for the Presence of Cyclostationarity (U. V.A. Report 192444B/EEE93/134, Dec. 1992).

[18] F. Heller, H.L. Hurd, T. Lindström, G. Söderbacka, Cyclostationarity and Periodically Perturbed Systems (In preparation).

SIGNALS ASSOCIATED WITH NONLINEAR DYNAMICAL SYSTEMS: IDENTIFICATION AND MONITORING

Jon Wright
Institute for Nonlinear Science
University of California at San Diego,
LaJolla, Ca 92093-0402

ABSTRACT

We discuss a method for analyzing signals from nonlinear dynamical systems. It is assumed that the signal is chaotic with a broad frequency spectrum, such as that associated with a strange attractor. The proposed method is useful for identifying a signal in the presence of other signals or noise. It is also useful for monitoring changes in signals in a noisy environment.

INTRODUCTION

A method is proposed for identification, classification and monitoring of changes of signals associated with nonlinear dynamical systems. It is intended for situations where several signals may be present, each from a different source (noise is just another signal). The particular signal of interest is assumed to be associated with a dynamical system having a low dimensional strange attractor. The frequency spectrum is presumed to have a broad band character, else the system could be analyzed with conventional linear signal processing tools. The goals are to either identify that a signal associated with a particular source is present, or to monitor the changes in a nonlinear system. One might for instance want to identify a sound as having a component coming from a ship of type 'A'. Or one might want to monitor the performance of a mechanical device.

Associated with any signal are an infinite number of joint probability distributions. Each probability distribution has its associated characteristic function, which is just the Fourier transform of the probability distribution. In nonlinear dynamics, these distributions are more commonly referred to as densities. The theory associated with nonlinear dynamics suggests that if a low dimensional strange attractor is present, then only low dimensional probability distributions are important. The higher order ones are in principle determined in terms of the low order ones. This is important, because otherwise one might have to consider very high dimensional densities which is computationally impractical.

In this paper, we will only present calculations using a particular class of one dimensional characteristic functions. For examples using the densities and higher dimensions, see my other articles [1, 2]. One dimensional characteristic functions are easy to calculate and provide very clear signatures.

FORMALISM

We now proceed to outline the assumptions and the method. For more details see [1]. It is assumed that the incoming signal, $S(t)$ is sampled at times $t_o, t_o + \delta t, t_o + 2\delta t \ldots$
$S_n \equiv S(t_o + n\delta t)$ for a total of N values with total time $T_o = N\delta t$.

The choice of the sampling time is a balance between too much data and too little information. If there is a typical period present, one needs to sample at a rate of perhaps ten times per period. Experience will eventually provide more guidance as to a good choice of δt.

It turns out to be useful to remove the mean from all signals and to normalize them to unit strength so that they satisfy the equations,

$$\sum_{i=1}^{N} S(t_n) = 0. \qquad (1)$$

$$\sum_{i=1}^{N} S^2(t_n) = 1. \qquad (2)$$

The reason for this is that the amplitude will only scale the argument of the characteristic function, and it is convenient to work with a fixed scale. The amplitude and mean can be followed separately. When the signal is the sum of signals from several sources, the relative amplitude is an important parameter.

The simplest density, which is not very useful for identification and monitoring, is given by

$$\rho(x) = \lim_{T_o \to \infty} \frac{1}{T_o} \int_0^{T_o} \delta(x - S(t))dt = \lim_{N \to \infty} \frac{1}{N} \sum_{j=1}^{N} \delta(x - S_j). \qquad (3)$$

It is just a histogram, i.e. $\rho(x)dx$ is the probability that a particular signal S_n is in the range x to $x + dx$ The delta function has to be replaced in practice by finite size bins or some smoothing function that is well localized at x. We choose to use the following approximation,

$$\delta(f) \to \frac{1}{\sigma\sqrt{2\pi}} e^{\frac{-f^2}{2\sigma^2}}. \qquad (4)$$

If we have only a scalar signal we can construct a second signal by using a time delayed signal as the second coordinate [3]. If we do that we get the following two-dimensional density,

$$\rho_T(x,y) = \lim_{T_o \to \infty} \frac{1}{T_o} \int_0^{T_o} \delta(x - S(t))\delta(y - S(t + T))dt, \qquad (5)$$

One can go to higher dimensions as well. The autocorrelation function of the signal is given by a second moment of the two dimensional density.

Signals Associated with Nonlinear Dynamical Systems

$$C(T) = \lim_{T_o \to \infty} \frac{1}{T_o} \int_0^{T_o} S(t)S(t+T)dt = \int xy\rho(x,y)dxdy \qquad (6)$$

Our normalization condition and removal of the mean impose the following constraints upon the densities.

$$\int x\rho(x)dx = \sum_{i=1}^{N} S(t_n) = 0. \qquad (7)$$

$$\int x^2 \rho(x)dx = 1. \qquad (8)$$

The characteristic function for a probability distribution is just the fourier transform of the density. It can be calculated directly from the signal as follows:

$$\hat{\rho}(k) = \int e^{ikx} \rho(x)dx = \lim_{T_o \to \infty} \frac{1}{T_o} \int_0^{T_o} e^{ikS(t)}dt = \lim_{N \to \infty} \frac{1}{N} \sum_{n=1}^{N} e^{ikS_n}. \qquad (9)$$

The usefulness of the characteristic function becomes very apparent when the sum of two signals is received. If the signals are labeled by 'a' and 'b' and the corresponding densities are constructed, then

$$S(t) = S_a(t) + S_b(t) \qquad (10)$$

and

$$\hat{\rho}_a(k) = \lim_{T_o \to \infty} \frac{1}{T_o} \int_0^{T_o} e^{ikS_a(t)}dt \qquad (11)$$

$$\hat{\rho}_b(k) = \lim_{T_o \to \infty} \frac{1}{T_o} \int_0^{T_o} e^{ikS_b(t)}dt. \qquad (12)$$

It is a well known mathematical theorem that the characteristic function for two independent processes is given by the product of the separate characteristic functions [4].

$$\hat{\rho}(k) = \hat{\rho}_a(k)\hat{\rho}_b(k). \qquad (13)$$

A similar relation holds for more signals, for example for three signals one has,

$$\hat{\rho}(k) = \hat{\rho}_a(k)\hat{\rho}_b(k)\hat{\rho}_c(k). \qquad (14)$$

Because we chose to normalize all signals, the actual relationships involve the relative normalizations,

$$S(t) = \alpha S_a(t) + \beta S_b(t) \tag{15}$$

$$\alpha^2 + \beta^2 = 1 \tag{16}$$

$$\hat{\rho}(k) = \hat{\rho}_a(\alpha k)\hat{\rho}_b(\beta k) \ . \tag{17}$$

It is also possible to derive a relationship where the relationship between two processes occurs as a sum rather than a product. The appropriate derivatives can be computed directly from the data, so it is not necessary to numerically differentiate anything.

$$\frac{d\hat{\rho}(k)}{dk} = \lim_{T_o \to \infty} \frac{1}{T_o} \int_0^{T_o} iS(t)e^{ikS(t)}dt. \tag{18}$$

$$\frac{1}{\hat{\rho}(k)}\frac{d\hat{\rho}(k)}{dk} = \frac{1}{\hat{\rho}_a(\alpha k)}\frac{d\hat{\rho}_a(\alpha k)}{dk} + \frac{1}{\hat{\rho}_b(\beta k)}\frac{d\hat{\rho}_b(\beta k)}{dk}. \tag{19}$$

It is useful for illustrative purposes to consider a simple example that can be done analytically. Suppose the incoming signal is harmonic.

$$S(t) = \alpha \cos(\omega t) \tag{20}$$

The density is easily found to be,

$$\rho(x) = \frac{1}{\pi\sqrt{(\alpha^2 - x^2)}}, -\alpha < x < \alpha \tag{21}$$

and the characteristic function turns out to be,

$$\hat{\rho}(k) = J_0(\alpha k) \tag{22}$$

Notice that only the amplitude information appears. The frequency information is not present. However if we go to two dimensions then we can obtain information about the frequency.

$$\hat{\rho}_T(k_1, k_2) = \lim_{T_o \to \infty} \frac{1}{T_o} \int_0^{T_o} e^{ik_1 \cos(\omega t) + ik_2 \cos(\omega t + \omega T)} dt \tag{23}$$

Now we don't actually have to go to two dimensions, as it is sufficient to consider a one dimensional slice through the two dimensional space. We choose the special case $k = k_1 = k_2$ and obtain,

$$\hat{\rho}_T(k, k) = J_0(2k\alpha \cos(\omega T/2)) \tag{24}$$

We could have also taken a one-dimensional slice through a three (or higher) dimensional space. This suggests that we consider one dimensional characteristic functions that are slices of higher dimensional ones. If necessary, we can construct two dimensional slices, but the examples that follow suggest that one dimensional functions may be sufficient in practice. We restrict ourselves for the present to the special one dimensional characteristic functions given by

$$\hat{\rho}_T(k) = \lim_{T_o \to \infty} \frac{1}{T_o} \int_0^{T_o} e^{ikS(t)) + ikS(t+T))} dt \qquad (25)$$

EXAMPLES

In order to illustrate the utility of such functions we present several examples. Since we are only demonstrating that the method works in principle, it is sufficient to consider a few simple examples. We will use two signals from electronic circuits [5], and one each from the Lorenz [6] and Rossler [7] equations. In the figures, the electronic circuits will be labeled by 'C' and 'H'.

For each signal and each pair of signals the characteristic function was constructed,

$$\hat{\rho}_T(k) = \lim_{T_o \to \infty} \frac{1}{T_o} \int_0^{T_o} e^{ikR(t)} dt \qquad (26)$$

where

$$R(t) = S(t) + S(t+T). \qquad (27)$$

For a pair of signals, the signal $S(t)$ is the result of combining two signals,

$$S(t, \mu) = S_1(t, \mu) + S_2(t). \qquad (28)$$

The parameter μ represents a system parameter which is used to simulate changes in a system. Figure 1 shows the characteristic function for all four signals with a time lag, $T = 3$. The main impression I want to convey with this figure is that each of the signals has a considerably different structure.

In the figure 2 the characteristic function for the signal 'H' is shown for four different time lags, T. The impression that you should retain from this figure is that there is a lot of structure in the variable T as well. If only one signal were present, these two figures suggest that it would be very easy to make an identification by using the characteristic function. As we will see below, even when there are multiple signals present, the characteristic function is a useful identifier.

It is interesting to relate whatever part of this structure that we can to ordinary spectral analysis. To do that it is useful to expand the log of the characteristic function in powers of k.

$$log(\hat{\rho}_T(k)) = 1. - \hat{\rho}_2 k^2 - i\hat{\rho}_3 k^3 + \hat{\rho}_4 k^4 \ldots \qquad (29)$$

The coeficients of the powers of k are given in terms of the cumulants of $R(t)$ [8]. The coefficient of k^2 is

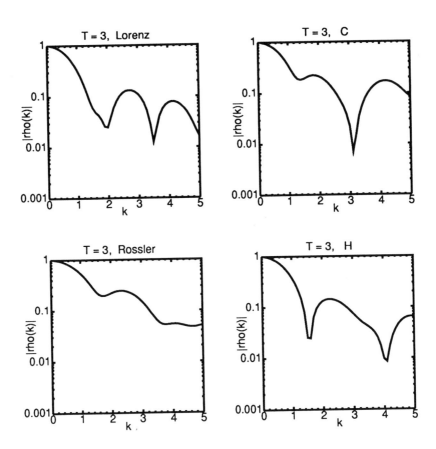

Figure 1: The characteristic function, $\hat{\rho}_T(k)$, for the four different signals at a time lag of $T = 3$. The signal source is indicated by the label at the top of each graph.

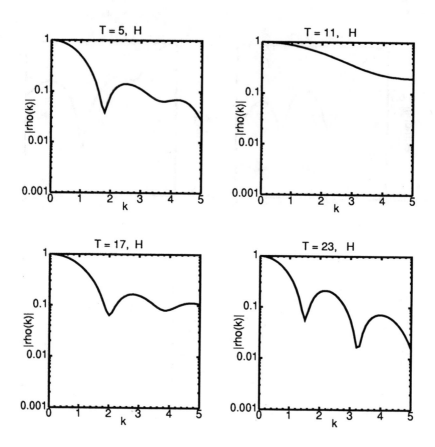

Figure 2: The characteristic function, $\hat{\rho}_T(k)$ for four time lags, $T = 3$, for the signal 'H'.

$$\hat{\rho}_2 = <(S(t)+S(t+T))^2>/2 = 1+C(T), \quad (30)$$

and the coefficient of $(ik)^3$ is

$$\hat{\rho}_3 = <(S(t)+S(t+T))^3>/6 \quad (31)$$

which is a combination of third moments of the signal.

The coefficient of k^4 is given as a combination of the second and fourth moments,

$$\hat{\rho}_4 = (<(S(t)+S(t+T))^4> -3<(S(t)+S(t+T))^2>^2)/24 \quad (32)$$

If the signal $S(t)$ were characterized by Gaussian statistics, as noise is often assumed to be the case, then the higher order cumulants vanish and its characteristic function would be given by

$$\hat{\rho}_T(k) = e^{-k^2(1+C(T))} \quad (33)$$

The part of the curve near $k = 0$ is determined by the auto correlation function or the spectrum. The next order contribution is determined by the third and fourth order moments. So the part of the curve near $k = 0$ is given entirely by the low order moments. The rest of the curve cannot in any reasonable way be determined from low order moments. This is a very important point and one should understand that the identification of a characteristic function by its entire curve is much easier than trying to use the first three or four terms in a power series around the origin, which is all that one can do with ordinary spectral and bi-spectral etc. calculations. In figure 3 the characteristic function of Eq. (23) for the Lorenz equations is shown for four different values of the time lag, T. The dotted curve is the approximation to the characteristic function that results from considering all moments of fourth order and lower. Clearly there is much more useful information in the rest of the curve than in the part determined by the low order moments. It should be mentioned that calculating the characteristic function for larger k values gets progressively more difficult as k increases, since usually the value of $\rho(\hat{k})$ for larger k values becomes small and a longer time series is needed in order to get statistically significant results.

We now wish to demonstrate the usefulness of the factorization property given in Eq. (17). As an example, suppose two signals of equal strength are added together and we wish to decide which signals are present. In figure 4a we show the characteristic function for the incoming signal, which in this case is 'H' plus Rossler for a time lag, $T = 5$. This is the left hand side of Eq. (17). In figures 4b-4d are shown the right hand side of the equation for three combinations of candidate signals. The particular signals are indicated

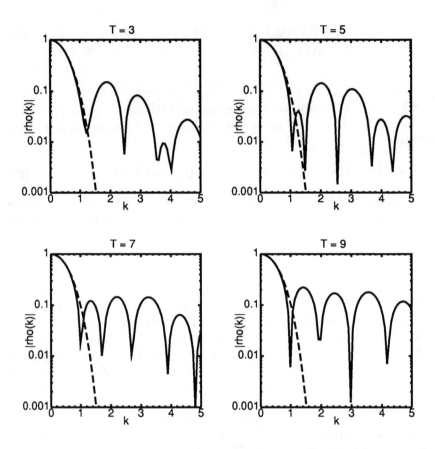

Figure 3: The solid line is the absolute value of the characteristic function, $\hat{\rho}_T(k)$ for four different values of the time lag T. The dotted line is the approximation to the same curve if only the first four moments of the time series are used.

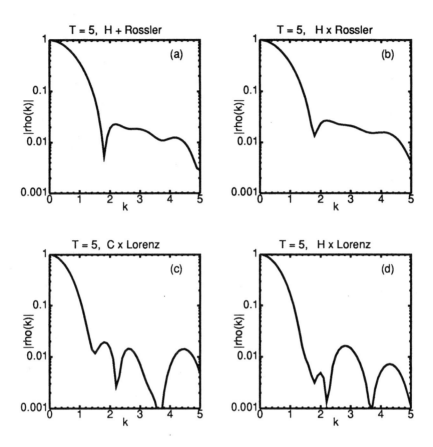

Figure 4: The left hand side of Eq. (17) is shown in (a) for the signal which is the sum of 'H' and Rossler. The right hand of the equation is shown in (b-d) for the signals indicated at the top of the figure. The time lag is $T = 5$

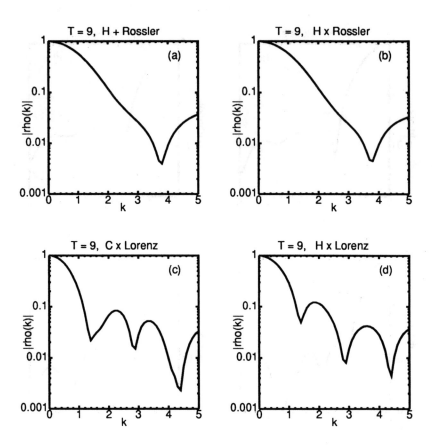

Figure 5: The left hand side of Eq. (17) is shown in (a) for the signal which is the sum of 'H' and Rossler. The right hand of the equation is shown in (b-d) for the signals indicated at the top of the figure. The time lag is $T = 9$

in the label above each graph. It is clear which is the correct combination of signals.

We can also consider other time lags in order to increase our confidence level. In figures 5a-5d the same combination of signals is shown, this time for a lag of $T = 9$. The fact that all graphs in figures 4 and 5 are significantly different, and that other time delays would give similar differences, should convince the reader that the characteristic functions considered in this paper provide a potentially powerful tool for signal identification.

Another potential application of these ideas is to the task of monitoring changes in nonlinear dynamical systems. For instance one might wish to detect deterioration in a mechanical system which could be an engine or a turbine blade, perhaps even an airplane wing. To illustrate the idea we again return to

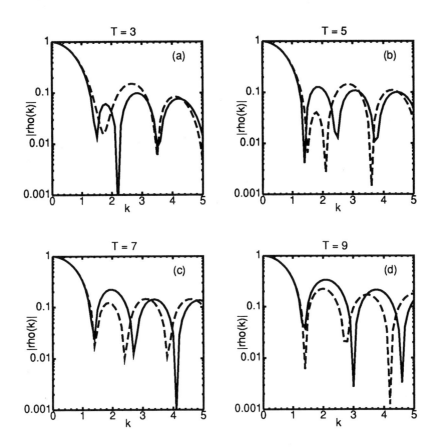

Figure 6: The absolute value of the characteristic function for a signal that is pure Lorenz. The solid and dashed curves are for two different values of a parameter in the Lorenz equations.

the Lorenz equations, and solve them for two different values of one of the parameters in the equations. The change in the parameter was sufficiently small so that there was neither an identifiable change in the signal time trace, nor a significant change in the spectrum. However the change in the characteristic function was easily observable. In figures 6a-6d two curves are shown for four different time lags. The solid line is the characteristic function for one value of the parameter, and the dotted line is for the second value of the parameter. There are many easily identifiable changes.

For a situation with a pure signal one can imagine that there may be many other ways of detecting changes. However in the presence of noise or other signals, the task becomes much more difficult. The method presented here still provides the same easily identifiable changes. To illustrate this point we

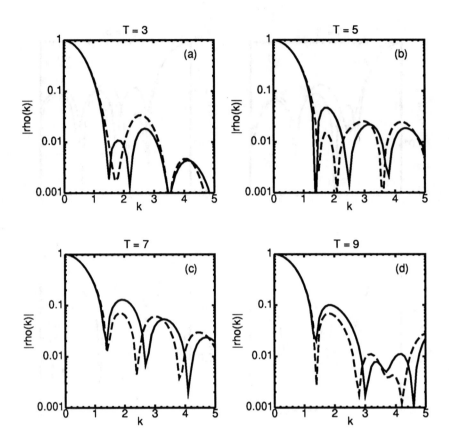

Figure 7: The absolute value of the characteristic function for a signal that is the sum of the Lorenz and Rossler signals. The solid and dashed curves are for two different values of a parameter in the Lorenz equations.

added equal amounts of Rossler plus Lorenz signals together for two different values of the parameter in the Lorenz equations. The results are shown in figure 7 and the same clear signatures are still present. It would be virtually hopeless to observe these changes by examining low order moments because of the presence of the Rossler signal. In order to make clear the relationship between the structures in the two figures, the k axis has been scaled by a factor of $\alpha = \sqrt{2}$ in figure 6 (see Eq. (17)).

REMARKS

One might reasonably ask why this method is not in wide use if it performs as suggested in this article. There are probably several reasons. First, by standards of a few years ago, considerable computational resources are required. However with the gains in computing capability of the last few years that is no longer a consideration. Even if it were for some application, the algorithms are an obvious candidate for parallel processing. Second, the last ten years has seen a considerable growth in our understanding of dynamical systems. The concept of strange attractors is new and that concept is accompanied by the realization that only a few degrees of freedom may be relevant in many dynamical systems. This means that it is not necessary to consider high dimensional densities because they are in principle determined in terms of lower dimensional densities if only a few degrees of freedom are relevant. This also implies that the lower dimensional characteristic functions or the associated densities can have sufficient structure to serve as identifiers. The concept of phase space reconstruction is also fairly new. Of course the densities we consider are really just multi-time joint probability densities, but the concept of phase space reconstruction suggested that it could be fruitful to consider them.

It would be reasonable to ask how low must the dimension be before the proposed algorithm would work well. We have done some very preliminary analysis of data for high Reynolds number flow past a body. The effective dimension was about eight, and yet the one dimensional characteristic function described in Eq. (9) still seemed to have significant structure. Perhaps there may be sufficient structure in the relevant slices through higher dimensional characteristic functions even for higher dimensional systems.

The issue of errors in the estimation of $\hat{\rho}(k)$ has not been addressed. Our only measure was to construct the function from the first and second half of a data set and compare for the electronic circuit data. In the case of the Lorenz and Rossler signals, we computed sufficiently long time series so that the fluctuations are unimportant for our purposes. The close agreement of figures 4a and 4b as well as 5a and 5b is a good indication that the fluctuations are small. As we remarked earlier, the characteristic function tends to be smaller for higher values of k and consequently has more fluctuations. In the example of flow noise mentioned in the paragraph above, the total time was about 5 seconds, and the two halves of the time series were in agreement for values of k less than about 3. Clearly the topic of error estimates needs further study.

SUMMARY

We have demonstrated that some one dimensional characteristic functions are very good identifiers of the source of a signal, and they work satisfactorily even in the presence of an equal amount of noise or other signal. The confidence of the estimation can be greatly improved by using a family of characteristic functions parameterized by the time lag T. These same functions also serve as excellent monitors of changes in dynamical systems that are not observable by spectral analysis. Although the methods would work in principle for any situation, it is expected that their greatest effectiveness will be in situations where the signal has a broad frequency spectrum and is associated with a low dimensional chaotic attractor so that conventional signal processing will not work very well.

References

[1] Jon Wright and Roy Schult, "Recognition and Classification of Nonlinear Chaotic Signals" preprint (1993) Institute for Nonlinear Science, University of California, San Diego. submitted for publication

[2] Jon Wright "Monitoring Changes in Time of Chaotic Nonlinear Systems" preprint (1993) Institute for Nonlinear Science, University of California, San Diego. submitted for publication

[3] J.-P. Eckmann and D. Ruelle, Rev. Mod. Phys. 57, 617-656 (1985) and references therein.

[4] A. Papoulis, Probability, Random Variables and Stochastic Processes (McGraw-Hill, 1991)

[5] The electronic circuit signals were provided by Tom Carroll and Lou Pecora of the Naval Research Laboratory.

[6] E. N. Lorenz, J. Atmos. Sci. 20, 130-41 (1963)

[7] O.E.Rossler, Phys. Lett. 57a, 397 (1976)

[8] N.G.Van Kampen, Stochastic Processes in Physics and Chemistry (North-Holland, 1981)

CHAOS IN PROPAGATION MODELING AND REAL ENVIRONMENTS

CHAOS IN AN ACOUSTIC PROPAGATION MODEL

Michael A. Wolfson and Frederick Tappert
Division of Applied Marine Physics
Rosenstiel School of Marine and Atmospheric Science
University of Miami
4600 Rickenbacker Cswy
Miami, FL 33149

ABSTRACT

A nonautonomous Hamiltonian dynamical system having one degree of freedom is derived physically by considering long range acoustic propagation in the presence of oceanic mesoscale structure in the classical limit (geometrical approximation). A stochastic Fokker-Planck theory is developed to analyze this nonintegrable dynamical system. The main result is that two particles (rays) that are initially separated by an infinitesimal amount diverge exponentially at a rate given by the Lyapunov exponent that has been calculated theoretically and compared to numerical experiments with agreement to two decimal places. The practical implication of this result is that deterministic ray tracing predictions in the ocean are not possible beyond the "predictability horizon" of a few thousand kilometers, due to horizontal plane multipaths induced by naturally occurring mesoscale activity.

INTRODUCTION

When the geometrical (ray) approximation is made, the linear acoustic wave equation is replaced by a system of nonlinear ordinary differential equations that can be cast in the form of a Hamiltonian dynamical system. When, in addition, the medium through which the sound waves propagate (the ocean) contains lateral inhomogeneities, then the ray equations are generically nonintegrable and ray trajectories exhibit chaotic behavior, i.e., extreme sensitivity to the initial conditions and/or the environment. As a consequence, accurate predictions of ray trajectories and arrival times are not possible beyond a certain "predictability horizon". This limitation of the ability to predict sound propagation is due ultimately to the finite precision of the measurement instruments that provide inputs to the propagation models.

The first published example of the above general observation concerning predictability of acoustic propagation was Palmer et al.[1] in the context of two-dimensional vertical plane (range, depth) propagation in the oceanic sound channel, perturbed by internal waves that were modeled as periodic in range. Analysis of chaotic behavior was based on Poincaré maps and numerically computed Lyapunov exponents that gave predictability horizons of a few hundred kilometers. A similar study that introduced area-preserving mappings was published by Brown et al.[2] Other studies of vertical plane propagation in shallow water with range-dependent bathymetric variations yielded predictability horizons of a few tens of kilometers [Tappert et al.,[3] Goñi[4]]. A summary of this research was published by Brown et al.[5] More recently, general methods for analyzing chaos in vertical plane oceanic sound channel propagation were discussed by Smith et al.,[6] and effects of mesoscale perturbations were numerically computed by Smith et al.[7] to yield predictability horizons of a few thousand kilometers.

The research reported in this article has two primary motivations: 1) to provide theoretical estimates of Lyapunov exponents and thereby predictability horizons, which heretofore had to be computed numerically; and 2) to estimate limitations of predictability for long-range low-frequency (LRLF) global ocean acoustic propagation, to support studies of acoustic monitoring of possible global climate change. Munk and Forbes[8] first proposed using LRLF acoustics to monitor possible global change induced by the greenhouse effect, and the Heard Island Feasibility Test (HIFT) was successfully carried out.[9] The current program, sponsored mainly by DARPA, is called ATOC (Acoustic Thermometry of Ocean Climate) and GAMOT (Global Acoustic Monitoring of Ocean Temperature).

In the second section of this article, an idealized LRLF two-dimensional horizontal plane acoustic propagation model is derived for each acoustic mode in depth. Mesoscale perturbations near the sound channel axis cause horizontal multipaths and acoustic chaos, which is rather different from the vertical plane chaos studied previously. In addition, the parabolic approximation and the geometric approximation are made, resulting in a nonautonomous Hamiltonian dynamical system having one degree of freedom.

In the third section, a stochastic Fokker-Planck theory is developed based on the method of smoothing, and analytical expressions are derived for momentum diffusion and for the expected value of the Lyapunov exponent.

The fourth section describes Monte-Carlo numerical simulation experiments of ray trajectories, and compares these results to the theory.

The final section summarizes the results of this research, describes the practical implications of the results, and states the conclusions.

DERIVATION OF DYNAMICAL SYSTEM EQUATIONS

We start with the three-dimensional parabolic wave equation (PE) in Cartesian coordinates in which x is range, y is cross-range, and z is depth. Ignoring earth curvature effects, we arrive at[10]

$$i\frac{\partial \Psi}{\partial x} + \frac{1}{2k_0}\left(\frac{\partial^2 \Psi}{\partial y^2} + \frac{\partial^2 \Psi}{\partial z^2}\right) - k_0 U(x,y,z)\Psi = 0 \;, \tag{1}$$

where $\Psi(x,y,z)$ is the normalized reduced acoustic pressure, and k_0 is the reference wave number. Here, the potential is expressed as

$$U(x,y,z) = \overline{U}(z) + \mu(x,y,z) \;, \tag{2}$$

where $\overline{U}(z)$ represents the average sound speed structure and $\mu(x,y,z)$ represents the mesoscale structure. The normal modes of the unperturbed, separable problem satisfy

$$\frac{d^2\phi_n}{dz^2} + 2k_0^2\left(\varepsilon_n - \overline{U}(z)\right)\phi_n = 0 \;, \tag{3}$$

where the ϕ_n are the orthonormal acoustic modes (i.e., $\int \phi_n(z)\phi_m(z)dz = \delta_{n,m}$). Expansion of the solution in terms of these normal modes,

$$\Psi(x,y,z) = \sum_n \psi_n(x,y)\phi_n(z)e^{-ik_0\varepsilon_n x} \;, \tag{4}$$

gives the coupled mode equations,

$$i\frac{\partial \psi_m}{\partial x} + \frac{1}{2k_0}\frac{\partial^2 \psi_m}{\partial y^2} - k_0 a_{mm}(x,y)\psi_m = k_0 \sum_{n\neq m} e^{ik_0(\varepsilon_m - \varepsilon_n)x} a_{nm}(x,y)\psi_n \;, \tag{5}$$

where

$$a_{nm}(x,y) = \int \phi_n(z)\mu(x,y,z)\phi_m(z)dz \;. \tag{6}$$

Although the mode coupling term on the right hand side of Eq.(6) is likely to be important in long range acoustic propagation, we will neglect it for the rest of this work. The results concerning chaos to be brought forth later in this paper are not invalidated by the neglect of mode coupling; mode coupling only adds unnecessary complications to the analytical treatment of the problem. Defining $\epsilon_m = \langle a_{mm}^2(x,y)\rangle^{1/2} \ll 1$, where the brackets denote spatial averaging, and $u_m(x,y) = a_{mm}(x,y)/\epsilon_m$, we obtain

$$i\frac{\partial \psi_m}{\partial x} + \frac{1}{2k_0}\frac{\partial^2 \psi_m}{\partial y^2} - k_0 \epsilon_m u_m(x,y)\psi_m = 0 \ . \tag{7}$$

Thus we have a two-dimensional horizontal parabolic wave equation for each acoustic mode. The potential, $u_m(x,y)$, is the integral over depth of the mesoscale structure weighted by the m'th acoustic normal mode, and finally normalized by its rms value. The rms value, denoted by ϵ, represents the strength of the sound speed fluctuations relative to some reference sound speed near the sound speed axis (i.e. $\epsilon \approx \langle (\delta c/c_0)^2\rangle^{1/2}$). In terms of modelling $u_m(x,y)$, we make the following gross simplifications: Given a choice of mode number m, we assume the potential has unit-variance and zero-mean, is spatially stationary, isotropic, and has a single scale length, $L = O(40km)$, and $10^{-3} \lesssim \epsilon \lesssim 10^{-2}$. For numerical simulations, we model $u(x,y)$ with a bivariate Gaussian spectrum,

$$S(k,l) = \frac{L^2}{\pi}e^{-L^2(k^2+l^2)} \ , \tag{8}$$

so that the spatial correlation function is also Gaussian:

$$\langle u(x,y)u(x+\xi,y+\eta)\rangle = \iint_{-\infty}^{\infty} S(k,l)e^{i(k\xi+l\eta)}dkdl \ ,$$
$$= \exp\left[-(\xi^2+\eta^2)/4L^2\right] \tag{9}$$

We now nondimensionalize Eq.(7) with the scale length L ($x \to x/L$, $y \to y/L$, $\delta = (k_0 L)^{-1}$). This gives, after dropping the modal subscript m,

$$i\delta\frac{\partial \psi}{\partial x} = -\delta^2/2\frac{\partial^2 \psi}{\partial y^2} + \epsilon u(x,y)\psi \ , \tag{10}$$

with the plane wave initial condition, $\psi(x=0,y)=1$.

The formal classical limit of Eq.(10), $\delta \to 0$, is readily obtained via the eikonal approximation. This gives the (parabolic) ray equations, equivalent to Newton's equations of motion in Hamiltonian form,

$$\frac{dy}{dx} = p \ ,$$
$$\frac{dp}{dx} = -\epsilon\frac{\partial u(x,y)}{\partial y} \ . \tag{11}$$

The plane wave initial condition is expressed as

$$y(0) = y_0 \ , \quad p(0) = 0 \ . \tag{12}$$

Since we will be interested in examining the dynamics of rays initially infinitesimally separated, we introduce the variational equations which correspond to the system just derived. Label the phase space variable by their initial crossrange position y_0:

$$y = y(y_0,x) \ , \quad p = p(y_0,x) \ . \tag{13}$$

Let

$$z = z(y_0,x) = \frac{\partial y(y_0,x)}{\partial y_0} \ , \quad \text{and} \quad w = \frac{\partial z}{\partial x} = \frac{\partial p(y_0,x)}{\partial y_0} \ . \tag{14}$$

Thus the variational equations are

$$\frac{\partial z}{\partial x} = w,$$
$$\frac{\partial w}{\partial x} = -\epsilon \frac{\partial^2 u}{\partial y^2} z, \quad (15)$$

with plane wave initial condition $z(y_0, 0) = 1$, $\dot{z}(0) = 0$. These variational equations play a major role in looking for chaos in the Hamiltonian ray dynamical system just derived, as we will see in the section describing the numerics.

FOKKER-PLANCK THEORY AND SOLUTIONS

The ray equations can be analyzed statistically by the method of smooth perturbations (MSP)[a good, complete description of the method is given by Frisch[11], and an illustration of it's usefulness was clearly demonstrated by Besieris and Tappert[12]]. The one particle (ray) distribution function, which we denote by $f(y, p, x)$, satisfies the stochastic Liouville equation,

$$\frac{\partial f}{\partial x} + p\frac{\partial f}{\partial x} - \epsilon \frac{\partial u}{\partial y}\frac{\partial f}{\partial p} = 0, \quad (16)$$

with plane wave initial data,

$$f(p, 0) = \delta(p). \quad (17)$$

In the context of applying MSP, we let $f = \langle f \rangle + \delta f$, where $\langle f \rangle$ represents the mean part of f, and δf represents the fluctuations about the mean. Taking the ensemble average of Eq.(16) yields

$$\frac{\partial \langle f \rangle}{\partial x} = -p\frac{\partial \langle f \rangle}{\partial y} + \epsilon \frac{\partial}{\partial p}\langle \frac{\partial u}{\partial y}\delta f \rangle. \quad (18)$$

Next we subtract this result from Eq.(16) to obtain (retaining only first order terms in δf)

$$\frac{\partial \delta f}{\partial x} = -p\frac{\partial \delta f}{\partial y} + \epsilon \frac{\partial}{\partial p}\frac{\partial u}{\partial y}\langle f \rangle. \quad (19)$$

We now solve Eq.(19) for δf and substitute this into Eq.(18). This yields

$$\frac{\partial \langle f \rangle}{\partial x} + p\frac{\partial \langle f \rangle}{\partial y} = \epsilon^2 \frac{\partial}{\partial p} \int_0^x dx' \langle \frac{\partial u(x,y)}{\partial y} \frac{\partial u(x+x', y+x'p)}{\partial y} \frac{\partial \langle f(y+x'p, p, x+x') \rangle}{\partial p} \rangle$$
$$\approx \epsilon^2 \frac{\partial^2 \langle f(p,x) \rangle}{\partial p^2} \int_0^x dx' \langle \frac{\partial u(x,y)}{\partial y}\frac{\partial u(x+x', y)}{\partial y} \rangle, \quad (20)$$

where in the last step we use the assumption that $\langle f \rangle$ varies slowly in range and crossrange. Also, to simplify the integral expression we assumed an isotropic power spectrum and used a small angle approximation, $|p| \ll 1$. Now we apply the long time Markovian approximation (LTMA)[12]. The spatial correlation in range of $\partial u/\partial y$ decays rapidly enough that, for $x \gg 1$, we may extend the limit of the integral in Eq.(20) to infinity. Since our initial condition is for plane waves and our power spectrum is isotropic, $\langle f \rangle$ must be independent of y and we finally obtain the diffusion equation for the mean distribution function:

$$\frac{\partial \langle f \rangle}{\partial x} = \epsilon^2 D_0 \frac{\partial^2 \langle f \rangle}{\partial p^2}, \quad (21)$$

with the plane wave initial condition

$$\langle f(p, x=0) \rangle = \delta(p). \quad (22)$$

The diffusion coefficient is an $O(1)$ parameter defined as

$$D_0 = \int_0^\infty dx' \langle \frac{\partial u(x,y)}{\partial y} \frac{\partial u(x+x',y)}{\partial y} \rangle , \qquad (23)$$

or, in terms of the power spectrum,

$$D_0 = \pi \int_{-\infty}^\infty S(0,l) l^2 dl . \qquad (24)$$

For the Gaussian power spectrum, we have

$$D_0 = \sqrt{\pi}/2 . \qquad (25)$$

The solution of Eq.(21) is

$$\langle f(p,x) \rangle = (4\pi\epsilon^2 D_0 x)^{-1/2} e^{-p^2/(4\epsilon^2 D_0 x)} . \qquad (26)$$

With this solution, all moments can be found. In particular,

$$\langle p^2 \rangle = 2\epsilon^2 D_0 x . \qquad (27)$$

This result was discovered by Chernov[13] and is in some sense uninteresting in that no extreme sensitivity is found. Thus we see that chaos cannot be discerned on the level of single particle statistics.

Next, we examine the relative motion of a pair of particles (rays) in the classical limit ($\delta = 0$), using the above ray equations and the method of smooth perturbations. We start with the stochastic Liouville equation for the 2-ray distribution function, $f_2(y_1, p_1, y_2, p_2, x)$,

$$\frac{\partial f_2}{\partial x} + p_1 \frac{\partial f_2}{\partial y_1} + p_2 \frac{\partial f_2}{\partial y_2} - \epsilon \frac{\partial u(x,y_1)}{\partial y_1} \frac{\partial f_2}{\partial p_1} - \epsilon \frac{\partial u(x,y_2)}{\partial y_2} \frac{\partial f_2}{\partial p_2} = 0 , \qquad (28)$$

subject to the plane wave initial distribution

$$f_2(y_1, p_1, y_2, p_2, 0) = \delta(y_1 - y_{10}) \delta(y_2 - y_{20}) \delta(p_1) \delta(p_2) \qquad (29)$$

For notational convenience we define the following operators:

$$\begin{aligned} L &= -p_1 \partial_{y_1} - p_2 \partial_{y_2} , & \text{(nonrandom)} , \\ \mathcal{L}(x) &= \epsilon \frac{\partial u(x,y_1)}{\partial y_1} \partial_{p_1} + \epsilon \frac{\partial u(x,y_2)}{\partial y_2} \partial_{p_2} , & \text{(centered random)} . \end{aligned} \qquad (30)$$

Equation (28) is now expressed as

$$\frac{\partial f_2}{\partial x} = (L + \mathcal{L}(x)) f_2 . \qquad (31)$$

Proceeding as before, we apply MSP and obtain the following closed system:

$$\frac{\partial \langle f_2 \rangle}{\partial x} = L \langle f_2 \rangle + \langle \mathcal{L}(x) \delta f_2 \rangle , \qquad (32)$$

$$\frac{\partial \delta f_2}{\partial x} = L \delta f_2 + \mathcal{L}(x) \langle f_2 \rangle . \qquad (33)$$

This system is readily solvable in closed form. Upon solving this system, and assuming that the mean pair distribution varies slowly in y_1, y_2 and x, and applying LTMA (just as for the one ray distribution), we arrive at (back in our original notation)

$$\frac{\partial \langle f_2 \rangle}{\partial x} + (p_1 \partial_{y_1} + p_2 \partial_{y_2}) \langle f_2 \rangle \approx \epsilon^2 \sum_{i,j} \frac{\partial}{\partial p_i} \left(D_{ij} \frac{\partial \langle f_2 \rangle}{\partial p_j} \right) \quad i,j = 1,2 , \qquad (34)$$

where the momentum ray diffusion coefficient is

$$D_{ij} = \int_0^\infty \langle \frac{\partial u(x,y_i)}{\partial y_i} \frac{\partial u(x-x', y_j - x' p_j)}{\partial y_j} \rangle dx' . \qquad (35)$$

We make a small angle approximation and the assumption that the power spectrum is isotropic to significantly simplify the expression for the diffusion coefficients. Defining the center of mass coordinates,

$$Y = \tfrac{1}{2}(y_1 + y_2), \quad y_r = y_1 - y_2 ,$$
$$P = \tfrac{1}{2}(p_1 + p_2), \quad p_r = p_1 - p_2 , \qquad (36)$$

we obtain

$$D_{11} = D_{22} = D_0 \qquad (37)$$

$$D_{12} = D_{21} = \int_0^\infty \langle \frac{\partial u(x,Y)}{\partial Y} \frac{\partial u(x-x', Y - y_r)}{\partial Y} \rangle dx' . \qquad (38)$$

Now we transform Eq.(34) to center of mass coordinates and integrate over the mean phase space Y, P to obtain

$$\frac{\partial \langle g \rangle}{\partial x} + p_r \frac{\partial \langle g \rangle}{\partial y_r} = \epsilon^2 \mathcal{D}(y_r) \frac{\partial^2 \langle g \rangle}{\partial p^2} , \qquad (39)$$
$$\langle g(y_r, p_r, x = 0) \rangle = \delta(p_r) .$$

Here $\langle g(y_r, p_r, x) \rangle = \iint dP dY \langle f_2(y_1, p_1, y_2, p_2, x) \rangle$ represents the mean relative pair distribution function, and $\mathcal{D}(y_r) = 2[D_0 - D(y_r)]$ is expressed in terms of the power spectrum as

$$\mathcal{D}(y_r) = 4\pi \int_{-\infty}^\infty S(0,l) l^2 \sin^2(l y_r / 2) dl . \qquad (40)$$

For large y_r, we have

$$\mathcal{D}(y_r) \to 2D_0 = \text{const} . \qquad (41)$$

Then the two rays diffuse independently of one another, and the solution of Eq.(39) is given by

$$\langle g(y_r, p_r, x) \rangle \to (8\pi \epsilon^2 D_0 x)^{-1/2} e^{-p_r^2/(8\epsilon^2 D_0 x)} . \qquad (42)$$

This expression may be used as the boundary conditions as $|y_r| \to \infty$ for the solution of Eq.(39). It may be interpreted in terms of the factorization:

$$\langle g(y_r, p_r, x) \rangle \to \int_{-\infty}^\infty \langle f(p_1, x) \rangle \langle f(p_2, x) \rangle dP . \qquad (43)$$

For small y_r, we have

$$\mathcal{D}(y_r) \to \alpha^2 y_r^2 , \qquad (44)$$

where

$$\alpha^2 = \pi \int_{-\infty}^\infty S(0,l) l^4 dl . \qquad (45)$$

For the Gaussian spectrum we have

$$\alpha^2 = 3\sqrt{\pi}/4 \,. \tag{46}$$

In fact, for all y_r we obtain

$$\mathcal{D}(y_r) = \sqrt{\pi}\left[1 - (1 - y_r^2/2)e^{-y_r^2/4}\right] \,. \tag{47}$$

The vanishing of $\mathcal{D}(y_r)$ at $y_r = 0$ means that two particles that initially have zero separation and relative momentum will always have zero separation and relative momentum, that is, the two particles will have the same phase space trajectories: $y_1(x) = y_2(x)$ and $p_1(x) = p_2(x)$. This solution is unstable, however, because a small (non-zero) initial separation of the particle positions causes exponentially increasing separations as range increases. This is the phenomenon of classical chaos.

The solution of Eq.(39) appears difficult to obtain analytically, except when $|y_r| \gg 1$. In terms of $\langle g(y_r, p_r, x) \rangle$, the intensity correlation function is given by

$$\langle I(Y + \tfrac{1}{2}y_r, x)I(Y - \tfrac{1}{2}y_r, x) \rangle = \langle h(y_r, x) \rangle$$
$$= \int_{-\infty}^{\infty} \langle g(y_r, p_r, x) \rangle dp_r \,, \tag{48}$$

where the intensity is $I(y_r, x) = |\psi(y_r, x)|^2$. For small y_r, Eq.(39) becomes

$$\frac{\partial \langle g \rangle}{\partial x} + p_r \frac{\partial \langle g \rangle}{\partial y_r} = \epsilon^2 \alpha^2 y_r^2 \frac{\partial^2 \langle g \rangle}{\partial p_r^2} \,. \tag{49}$$

This expansion is uniformly valid only when the separation y_r is initially infinitesimal, for otherwise it rapidly grows to become $O(1)$. For infinitesimal y_{r0}, a closed system of ordinary differential equations can readily be derived:

$$\frac{d\langle p_r^2 \rangle}{dx} = 2\epsilon^2 \alpha^2 \langle y_r^2 \rangle \,, \tag{50}$$

$$\frac{d\langle y_r^2 \rangle}{dx} = 2\langle y_r p_r \rangle \,, \tag{51}$$

$$\frac{d\langle y_r p_r \rangle}{dx} = \langle p_r^2 \rangle \,. \tag{52}$$

The solution of this system yields expressions for the second and cross moments of the variational quantities defined in the previous section. We find

$$\sigma_z^2 = \lim_{y_{r0} \to 0} \sigma_{y_r}^2/y_{r0}^2 = \frac{1}{3}\left[e^{2\nu_0 x} + 2\cos(\sqrt{3}\nu_0 x)e^{-\nu_0 x} - 3\right] \,, \tag{53}$$

$$\sigma_w^2 = \lim_{y_{r0} \to 0} \sigma_{p_r}^2/y_{r0}^2 = \frac{2}{3}\nu_0^2 \left\{e^{2\nu_0 x} - \left[\cos(\sqrt{3}\nu_0 x) - \sqrt{3}\sin(\sqrt{3}\nu_0 x)\right]e^{-\nu_0 x}\right\} \,, \tag{54}$$

$$\langle zw \rangle = \lim_{y_{r0} \to 0} \langle y_r p_r \rangle/y_{r0}^2 = \frac{1}{3}\nu_0 \left\{e^{2\nu_0 x} - \left[\cos(\sqrt{3}\nu_0 x) + \sqrt{3}\sin(\sqrt{3}\nu_0 x)\right]e^{-\nu_0 x}\right\} \,, \tag{55}$$

where the Lyapunov exponent ν_0 is given by

$$\nu_0 = \epsilon^{2/3}(\alpha^2/2)^{1/3} \,. \tag{56}$$

From Eq.(46) for a Gaussian spectrum, we obtain

$$\nu_0 = \epsilon^{2/3} A \,, \tag{57}$$
$$A = (3\sqrt{\pi})^{1/3}/2 \approx 0.87271... , \tag{58}$$

and from Eq.(53) we have

$$\sigma_z \approx \begin{cases} \sqrt{\frac{4}{3}} (\nu_0 x)^{3/2}, & \nu_0 x \ll 1 \\ \frac{1}{\sqrt{3}} e^{\nu_0 x}, & \nu_0 x \gg 1 \end{cases} \quad (59)$$

The large $\nu_0 x$ behavior shows the exponential growth characteristic of classical chaos. We note that the expression given in Eq.(53) was obtained by Malakhov et al. in 1977.[14] Also, a more rigorous derivation of Eq.(39), based on the probabilistic theorem of Papanicolaou and Kohler, was carried out by Zwillinger and White[15], who also arrived at the result in Eq.(53).

NUMERICAL EXPERIMENTS, COMPARISON TO THEORY

The numerical model consists of two parts: 1) the generation of a realization of mesoscale fluctuations on the axis; and 2) the solution of Eq.(11) and Eq.(15). Also, in order to illustrate the proliferation of caustics, we integrate the Lagrangian to obtain the eikonal, s, related to the travel time for the rays.

The method used to generate the mesoscale fluctuations is based on the two dimensional convolution of zero mean, unit variance, independent random phases with the power spectrum of the medium, in this case a bivariate Gaussian function. The real part of the two dimensional fast Fourier transform (FFT) of this convolution then yields a single realization of the medium. To generate the random phases we used a pseudo-random number generator based on a linear congruential method[16] which supplied uniformly distributed numbers on the unit interval. A Box-Muller transformation was then applied to yield Gaussian random numbers on the domain of reals with zero mean and unit variance. This spectral representation method of simulating random media is quite common, and a good review of its implementation and characteristics can be found in Shinozuka and Deodatis.[17] Note that we model the medium with periodic

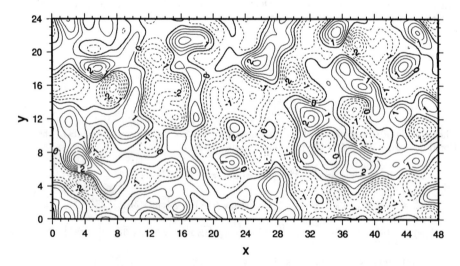

FIG. 1. A single realization of the mesoscale structure near the sound speed axis used for numerical simulations. The level one contours represent 7.5m/s fluctuations in sound speed (using $\epsilon = 5 \times 10^{-3}$). Both range and crossrange are in units of the correlation length scale of the eddies, $L = O(40\text{km})$.

boundaries in both range and crossrange; thus the number of degrees of freedom actualized in the statistical averaging will be limited by the periods chosen in relation to the correlation length scale L. We used a Fourier transform method to calculate both $\partial u/\partial y$ and $\partial^2 u/\partial y^2$. To obtain a medium with the desired statistical properties, it is necessary to use on the order of ten mesh points per physical unit. The size of the medium was based on RAM constraints of the Sun Spark II workstation used (32 MBs). We chose a size of 96 × 96 physical units with 1024 × 1024 mesh points. A section of a single realization is illustrated in Fig. 1.

A fourth order Runge-Kutta method was used to integrate the system

$$\dot{y} = p, \quad y(0) = y_0,$$
$$\dot{p} = -\epsilon \frac{\partial u}{\partial y}, \quad p(0) = 0,$$
$$\dot{s} = p^2/2 - \epsilon u, \quad s(0) = 0,$$
$$\dot{z} = w, \quad z(0) = 1,$$
$$\dot{w} = -\epsilon \frac{\partial^2 u}{\partial y^2} z, \quad w(0) = 0.$$

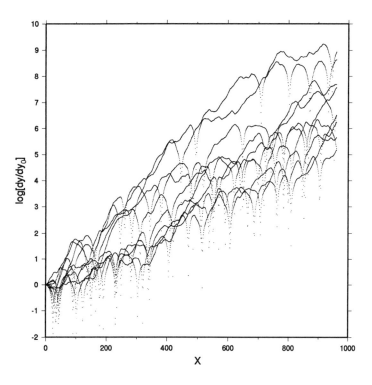

FIG. 2. Ten sample paths of the variation in crossrange, $z = \frac{\partial y}{\partial y_0}$; the nulls represent zero crossings, where caustics of the phase front occur.

The initial condition for a plane wave was represented by uniformly distributing 1500 rays, each with zero initial momentum along the extent of the crossrange axis. We used a bilinear interpolation scheme to evaluate the medium quantities between their computational mesh points. The range step used for integrating the rays was $\Delta x = 0.09375$. A natural connection between the Lyapunov exponent derived in the previous section to the variation equations for our system dictates that we define

$$\nu = \lim_{x \to \infty} x^{-1} \ln \left\langle \left(\frac{\partial y}{\partial y_0} \right)^2 \right\rangle^{1/2}. \tag{60}$$

The complex behavior of trajectories of $\partial y / \partial y_0$ shown in Fig. 2 demonstrates the need to compute the expected value of the Lyapunov exponent. The quantity in Eq.(60) is what was numerically computed, but the first order method of smoothing used in the previous section is only exact in the limit of a delta correlated medium, so we computed ν for different values of ϵ and extrapolated to $\epsilon = 0$ (See Fig. 3). We applied the ergodic hypothesis when computing $\nu(\epsilon)$, meaning the ensemble average in Eq.(60) was taken not only over separate realizations, but also spatially averaged using all the initially uniformly distributed rays. For $\epsilon = 5 \times 10^{-4}$, a least squares fit for $x > 300$ yielded $\nu/\epsilon^{2/3} = 0.8694$, which is within 0.38% of the theoretical value of A given by Eq.(58).

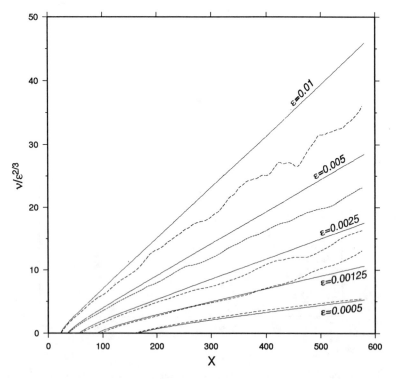

FIG. 3. Comparison of Eq. (53) with the numerically computed estimates of σ_z^2 for various values of ϵ. Estimates of ν were obtained by fitting the computed σ_z^2 curves to straight lines for $\epsilon^{2/3} x \gg 1$, and were scaled by $\epsilon^{2/3}$ in order to compare to the analytical estimate $\nu_0 = \lim_{\epsilon \to 0} \nu$.

CONCLUSION

Although we do not expect to observe deterministic chaos in the real ocean acoustic environment (the definition necessitates an infinite limit in range and infinitesimal perturbations), the divergences found above lead us to believe that signal predictability will be highly degraded over sufficiently large propagation distances due solely to horizontal mesoscale activity. Indeed, for typical horizontal mesoscale ($L = O(40\text{km})$ and $\epsilon = 5 \times 10^{-3}$), the predictability horizon, the distance at which the proliferation of multiple arrivals makes the signal unresolvable, is $O(1500km)$, which is on the scale of ATOC experiments. The fact that we neglected mode coupling indicates that our predictability horizon estimate is an upper bound, i.e., mode coupling only complicates matters and will further decrease the estimate.

One way to interpret these results is to take for granted that we know our environment (realization of the ocean on the mean sound channel axis depth) exactly, but our transmitter position is in error. In this sense we are concerned with a deterministic dynamical system whose solution depends nonlinearly on the environment. This small error in the transmitter position evolves to a large mean square cross-range separation distance to where the ray path should have gone. Another way to interpret these results is to surrender to the notion that we will never know the environment accurately enough to predict the acoustic signal at global scales of propagation.

All the above stated results depend on the strict classical limit, $\delta \rightarrow 0$, or frequency becoming infinite. Global acoustic measurements are made at finite frequency (about 100 Hz), and then $\delta \approx 10^{-4}$ which is very small but not zero. The study of finite frequency effects must be based on the linear wave equation, Eq.(10). Our numerical and theoretical (using MSP) studies of Eq.(10) have not revealed any sign of exponential sensitivity to either initial or environmental perturbations, not even as a transient condition that saturates at long range. Thus our current view of chaos in ocean acoustic propagation is that chaotic (exponential) behavior is an artifact of the classical limit. This implies that acoustic signals are in fact predictable at global ranges, but not by ray-based methods. Inversion of acoustic signals to obtain mesoscale structure is therefore possible if full-wave methods are used, but is not possible using ray-based methods, due to "ray chaos".

ACKNOWLEDGEMENTS

This work was supported by the Office of Naval Research and the National Science Foundation.

---

[1] D. R. Palmer, M. G. Brown, F. D. Tappert, and H. F. Bezdek, Geophys. Res. Lett. **15**, 569 (1988).
[2] M. G. Brown, F. D. Tappert, and G. Goni, Wave Motion **14**, 93 (1991).
[3] F. Tappert, M. Brown, and G. Goni, Phys. Lett A **153**, 181 (1991).
[4] G. Goni, Ph.D. thesis, University of Miami, 1991.
[5] M. G. Brown, F. D. Tappert, G. J. Goni, and K. B. Smith, in *Ocean Variability and Acoustic Propagation*, edited by J. Potter and A. Warn-Varnas (Kluwer Academic Publishers, Dordrect, Holland, 1991), pp. 139-160.
[6] K. B. Smith, M. G. Brown, and F. D. Tappert, J. Acoust. Soc. Am. **91**, 1939 (1992).
[7] K. B. Smith, M. G. Brown, and F. D. Tappert, J. Acoust. Soc. Am. **91**, 1950 (1992).
[8] W. H. Munk and A. M. G. Forbes, J. Phys. Ocean. **19**, 1765 (1989).
[9] A. Baggeroer and W. H. Munk, Phys. Today **45**, 22 (1992).
[10] F. D. Tappert, in *Wave Propagation and Underwater Acoustics*, edited by J. B. Keller and J. S. Papadakis (Springer-Verlag, New York, 1977), pp. 224-287.

[11] U. Frisch, in *Probabilistic Methods in Applied Mathematics*, edited by A. T. Bharucha-Reid (Academic Press, New York, New York, 1968), Vol. 1, pp. 114–122.
[12] I. Besieris and F. Tappert, J. Math. Phys. **17**, 734 (1975).
[13] L. A. Chernov, *Wave Propagation in a Random Medium* (McGraw-Hill, New York, 1960).
[14] A. N. Malakhov, S. N. Molodtsov, and A. I. Saichev, Radiophysics and Quantum Electronics **20**, 169 (1977).
[15] D. I. Zwillinger and B. S. White, Wave Motion **7**, 207 (1985).
[16] D. E. Knuth, *The Art of Computer Programming* (Addison-Wesley, Reading, Massachusetts, 1969).
[17] M. Shinozuka and G. Deodatis, Appl. Mech. Rev. **44**, 191 (1991).

DIMENSIONALITY, PREDICTION, AND DETERMINISM IN THE ANALYSIS OF REAL BROADBAND DATA

Richard F. Wayland, Jr., David Bromley, Douglas Pickett, Mary Eileen Farrell, and Anthony Passamante

Naval Air Warfare Center
Warminster, Pennsylvania 18974

ABSTRACT

In this paper new methods of nonlinear analysis are presented that provide a means of characterizing real world broadband signals in a noisy environment. In particular, evidence of *low-dimensionality* and *predictability* in these signals, reconstructed in the phase space, constitute strong indicators pointing to the nonlinear/chaotic nature of these signals. The nonlinear information captured from these signals may then be exploited for signal processing gains.

INTRODUCTION

NAWCAD Warminster has been actively involved in the application of nonlinear dynamics and chaos theory to the analysis of real world broadband signals in a noisy environment. Early on, NAWCADWAR attempted to apply some conventional methods of nonlinear dynamics, such as the calculation of Lyapunov exponents[1] and the correlation dimension,[2] to the analysis of real data, but those attempts met with no success.

Subsequently, research was initiated to develop new noise resistant and robust techniques to quantify and measure a potentially chaotic process. For this reason, the method based on the Local Intrinsic Dimension (LID),[3,4] was formulated and applied to real data with some success.

The results of the LID application have demonstrated that pertinent real broadband signals exhibit a low-dimensional character in the phase space domain. Low dimension in the phase space is an indicator of the data's complexity; in a sense they are proportional. These real data results were the first substantial indication that experimental broadband data could be so characterized.

The work that is discussed in this paper continues in the quest to describe as much as possible about real broadband data from a nonlinear dynamics point of view. This work and the results presented herein represent an effort to deal with noisy chaotic data by suppressing, modeling, and reorganizing that data in the phase space domain. To accomplish this task, methods (e.g., the *translation method*), were developed to model the data using the structure and phase space flow properties of the data available.

The major assumption for the *translation method* is that the local flow of the data in the phase space can be modeled by a linear transformation. Using this assumption, groups of clustered points in the phase space may be used to predict some points a short distance ahead in time. Then this prediction may be used as a criterion to make judgments as to whether the data, in any way, conforms to the modeling procedure and, therefore, whether it is deterministic in nature. This is where the tenuous connection between predictability and determinism is made.

Therefore, the quantity of interest is the signal prediction, or the error between the actual data and the prediction of that data. This idea comes from the fact that the prediction error will be smaller for data that conforms to the model used to make the prediction, than for noise or a noisy process. As will be seen below a calculated measure (the correlation coefficient) is made which indicates how close the predicted time series is to the data.

The ability to make predictions using the approaches outlined in this report is, in itself, a stronger indicator of a potentially chaotic process, than just that of the low dimension produced by the correlation dimension or the LID. This work was initiated to corroborate and strengthen earlier findings with regards to the low-dimensionality of real broadband data. Furthermore, these results can provide a new point of departure for achieving signal processing gains.

THEORY

The methods discussed below are aimed at the practical and robust characterization of real data from a nonlinear dynamics/chaos viewpoint. The processing is significantly different from any conventional approach in that the processing begins by embedding the data in the so-called phase space.

LOCAL INTRINSIC DIMENSION METHOD

This approach begins by using sequential time data $x(t)$ (suitably digitized and preprocessed to the desired bandwidth), such that
$$x(t) = \{x(1), x(2), \ldots x(n)\},$$
and $x(t)$ is embedded in an E-dimensional phase space by forming vectors (v_i) containing E elements, by grouping contiguous time points, E at a time, while sliding along

the data, skipping L data points (commonly known as the lag) between elements of each group as follows:

$$v_1 = [x(1),x(1+L),x(1+2L),\ldots,x(1+(E-1)L)]^T$$
$$v_2 = [x(2),x(2+L),x(2+2L),\ldots,x(2+(E-1)L)]^T$$
$$\vdots \qquad (1)$$
$$v_k = [x(k),x(k+L),x(k+2L),\ldots,x(k+(E-1)L)]^T$$

where T denotes the transpose. When time data that is generated by a deterministic system is reformatted or embedded in this manner, it will produce a well-defined geometric structure in the phase space called an attractor. A sufficient number of data samples, such as 20,000, is used to produce an accurate attractor portrait.

The essential idea in the Local Intrinsic Dimension approach is to randomly divide the entire attractor into a finite number of small local regions, each of which has a characteristic spatial distribution of data points extending locally into various orthogonal dimensions (as illustrated below).

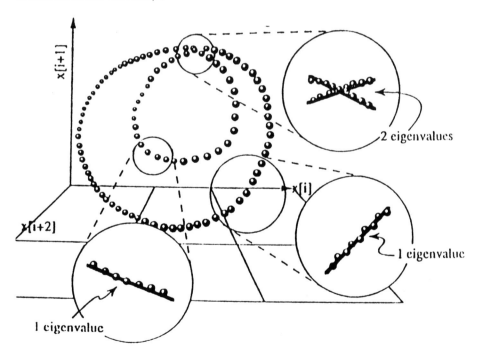

The number of these local orthogonal dimensions is the Local Intrinsic Dimension (LID). It is determined by the rank of the local data matrix X formed from the q nearest neighbors (after subtracting out the local center) of each local region, $X = (v_1, v_2, \ldots v_q)^T$, where q typically ranges between 30 and 60. The rank of X is determined using conventional singular value decomposition (SVD). In general, the output of the SVD contains E singular values, some signal and some noise, and the eigenvalues of the estimated data covariance matrix $R_x = X^T X$ are the squares of the singular values. A thresholding mechanism, such as the 3 dB criterion (i.e., a 50% drop in ratio of the successive eigenvalues), is then applied to separate the signal eigenvalues from the noise eigenvalues. The eigenvalues so produced will be indicative of the number of directions along which the data samples are distributed in a given local region. Since the LID changes with position on the attractor, a weighted average of the LID values is calculated to give the average dimension of the overall attractor.

PHASE SPACE PREDICTION--TRANSLATION METHOD

In addition to dimension estimation, the local flow properties of the data reconstructed in the phase space domain are investigated. For a chaotic process, the vector points within a local region probably will not be vector points that are sequential in the time domain. However, for some initial local region, if each of the nearest neighbors in the phase space domain are located, the next sequential vector point in the time domain (i.e., one step ahead in time, for all those nearest neighbors in the initial local region) may also be located. Each of the "one-step-ahead" vector time points that follow the initially selected vector time points will also fall into a local region, the absolute location of which may not be near the original local region in phase space.

However, because all the vector time points are only one time increment ahead, in this work the phase space transformation from one local region to the local region containing the next vector points in time, as discussed above, will be assumed to be given by a linear mapping. Actually, this is equivalent to an operation in the time domain for segments of data that are disjoint in time but close to each other in value.

Mathematically, the above discussion can be formulated as the following problem: assume that there is a smooth map f such that

$$v(j) = f(v(j-1)). \qquad (2)$$

Develop an approximation for f in a neighborhood of phase space containing the observed sequence of vectors v(1), ..., v(N).

Although a number of methods for approximating f have been investigated,[5-8] the most successful of these have been based on *local approximations*. That is, since f is a smooth function it can be approximated by simple constant or linear functions over small regions of phase space. Although the parameters in these *local* approximations will vary from one place on the attractor to another, they can be "pieced together" to form a global approximation to f. A specific approximation may now be developed based on the assumption that f is locally a translation.

Let v be a point close to the experimental attractor formed by v(1), ..., v(N). The idea is to approximate f(v). Toward this end, let v_1, ..., v_k denote the k nearest neighbors of v, and let w_1, ..., w_k denote their respective images under f (as shown below for k=5).

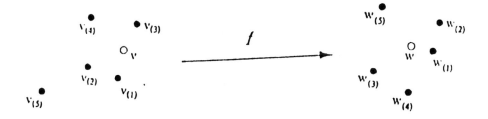

Assume that in a neighborhood of v containing v_1, ..., v_k one can write

$$f(y) \approx y + b, \qquad (3)$$

for some translation vector b. (For more information on the accuracy of this approximation see Refs. 9 and 10.) Using the kth nearest neighbor information described above, the least squares solution for b is

$$b = \frac{1}{k}\sum_{i=1}^{k}[w_i - v_i]. \qquad (4)$$

Eq. (4) is used to define the *local translation approximation* for f(v):

$$f_{trans}(v) = v + \frac{1}{k}\sum_{i=1}^{k}[w_i - v_i]. \qquad (5)$$

According to (5), $f_{trans}(v)$ is obtained by translating v through the average displacement experienced by its k nearest neighbors. Although (5) is a crude approximation to the map f, it is sufficiently accurate for the noisy nonstationary type of real acoustic data being analyzed here.

Although approximate forms for f can be used for many purposes, we focus on using f_{trans} to measure the phase space predictability of the time series $x(t)$. The procedure employed is as follows: select a time window of 2W points, say $x(1)$, ..., $x(W)$, ..., $x(2W)$ from the time series $x(t)$. Construct a "history attractor" from the first W points of the sequence, and a "future attractor" from the last W points of the sequence--this is done as prescribed earlier. The history attractor is then used to predict the future attractor. Specifically, the image of each point on the future attractor is predicted with f_{trans} using neighbors from the history attractor. The outcome of this procedure is a sequence of predicted values $\hat{x}(W+(E-1)L+2),...,\hat{x}(2W)$, corresponding to the observed values $x(W+(E-1)L+2),...,x(2W)$.

As a quantitative measure of the outcome of this prediction process, following Sugihara and May,[7] the *correlation coefficient rho* is used between the predicted time series and the time series actually observed. Specifically, *rho* is given by

$$\rho = \frac{\frac{1}{N_{preds}}\sum_{i=1}^{N_{preds}}(x_{W+(E-1)L+i}-<x>)(\hat{x}_{W+(E-1)L+i}-<\hat{x}>)}{\sqrt{\frac{1}{N_{preds}}\sum_{i=1}^{N_{preds}}(x_{W+(E-1)L+i}-<x>)^2}\sqrt{\frac{1}{N_{preds}}\sum_{i=1}^{N_{preds}}(\hat{x}_{W+(E-1)L+i}-<\hat{x}>)^2}}, \qquad (6)$$

where $<x>$ and $<\hat{x}>$ represent the average values of the observed and predicted sequences $x_{W+(E-1)L+i+1}$ and $\hat{x}_{W+(E-1)L+i+1}$, respectively, and $N_{preds}=2W-(E-1)L$. It easily follows that $-1 \leq \rho \leq 1$. Values of rho close to 1 indicate very good predictability, while values of rho close to 0 indicate poor predictability.

By sliding the observation window throughout the observed time series $x(t)$, a measure of the phase space predictability of $x(t)$ is obtained as a function of time. Of particular interest are the differences in rho between windows containing "signals of interest and background noise", and those containing merely "background noise".

NUMERICAL RESULTS

The above techniques were applied to several real broadband data sets, selected and prefiltered so that certain power spectra were isolated for simplicity. There are no restrictions on the signals to be analyzed except that, since composite signals are not yet of interest, they should be emanating from the same source, i.e., from the same actual process.

Fig. 1 shows the result of the application of the LID algorithm to a typical real data set. Clearly, the data exhibits a low dimension in the vicinity of the signal occurrence (corresponding to the first energy peak) and a high dimension in the region before and after the signal occurrence where only ambient noise is present. As a second measure of the potentially deterministic nature of this signal, the correlation coefficient rho, which gives a measure of the degree of predictability, is also computed. A gram-like display, called the *rho-gram*, giving rho as a function of time and number of predictions ahead, is also shown in Fig. 1. Notice that rho takes on a high value in the neighborhood of the signal occurrence and a low value in the regions before and after. This implies a high degree of predictability for this signal. In regards to the persistence of predictability, the rho value remains high for an extended number of points ahead, thus indicating a stable type of behavior. The energy statistic, however, is intrinsically unable to recognize underlying deterministic structure in a signal, and will therefore contribute, regardless of the type of source emanating the signal. This will serve to increase the effective false alarm rate. In the comparisons made, the LID method and the translation method both serve as natural discriminants of chaotic signals vs ambient noise.

CONCLUSIONS

The nonlinear methods presented in this paper offer new insight into the deterministic character of real broadband signals in noise. The properties of these nonlinear signals in the phase space, in particular, the flow of the trajectories on the attractor, may be exploited for signal processing gains. Generally, low LID values and high rho values indicate the presence of a low-dimensional signal, as well as a high probability that the signal is deterministic, since it is indicative of a low prediction error. More specifically, LID is generally low whenever signal-related energy is significant; rho is generally high at times corresponding to the highest signal energy but low before and after; and rho does not, in general, contribute in

296 Analysis of Real Broadband Data

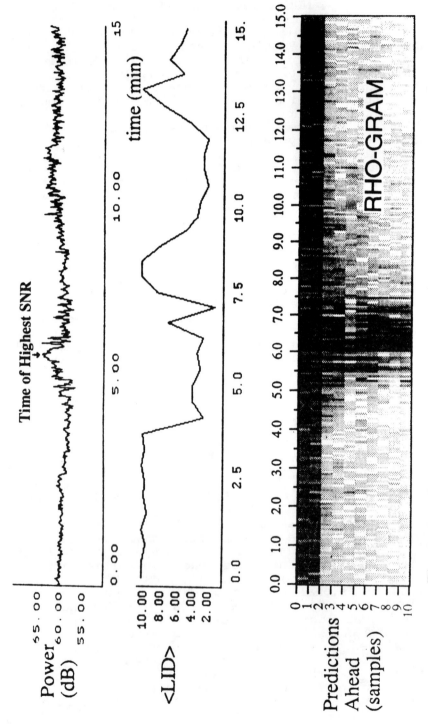

Fig. 1. Power, <LID>, and Rho for real broadband data.

the region of spurious energy spikes (as does the energy statistic).

ACKNOWLEDGMENTS

The authors would like to thank Tommy Goldsberry (ONR 231) and Dr. Kam Ng (ONR 122) for sponsoring this work.

REFERENCES

1. V. I. Oseledec, Trans. Moscow Math. Soc. $\underline{19}$, 197 (1968).
2. P. Grassberger and I. Procaccia, Physica $\underline{D9}$, 189 (1983).
3. A. Passamante, T. Hediger, and M. Gollub, Phys. Rev. $\underline{A39}$, 3640 (1989).
4. T. Hediger, A. Passamante, and Mary Eileen Farrell, Phys. Rev. $\underline{A41}$, 5325 (1990).
5. J. D. Farmer and J. J. Sidorowich, Phys. Rev. Lett. $\underline{59}$, 845 (1987).
6. M. Casdagli, Physica $\underline{D35}$, 335 (1989).
7. G. Sugihara and R. M. May, Nature $\underline{344}$, 734 (1990).
8. P. S. Linsay, Phys. Lett. $\underline{A153}$, 353 (1991).
9. R. Wayland, D. Bromley, D. Pickett, and A. Passamante, Phys. Rev. Lett. $\underline{70}$, 500 (1992).
10. R. Wayland, D. Pickett, D. Bromley, and A. Passamante, accepted for publication in the International Journal of Bifurcation and Chaos (1993).

Author Index

A

Abarbanel, H. D. I., 55

B

Bromley, D., 289
Broschart, T., 27
Brush, J. S., 159, 205

C

Carroll, T. L., 127
Carter, P. H., 233
Cawley, R., 14, 182, 193
Cembrola, J., 55

D

Ditto, W. L., 137

F

Farrell, M. E., 289
Frison, T., 43, 55

G

Galib, T., 55, 106

H

Hsu, G.-H., 182, 193
Hurd, H. L., 246

J

Jones, C. H., 246
Jones, C. K. R. T., 3

K

Kadtke, J. B., 159, 205
Katz, R. A., 55

M

Moon, F. C., 27

P

Passamante, A., 289
Pecora, L. M., 127
Pickett, D., 289

S

Salisbury, J., 106
Salvino, L. W., 14, 182
Spano, M. L., 137
Sreenivasan, K. R., 97
Stolovitzky, G., 97

T

Tappert, F., 277

W

Wayland, R. F., 289
Wolfson, M. A., 277
Wright, J., 260

AIP Conference Proceedings

		L.C. Number	ISBN
No. 179	The Michelson Era in American Science: 1870–1930 (Cleveland, OH, 1987)	88-83369	0-88318-379-X
No. 180	Frontiers in Science: International Symposium (Urbana, IL, 1987)	88-83526	0-88318-380-3
No. 181	Muon-Catalyzed Fusion (Sanibel Island, FL, 1988)	88-83636	0-88318-381-1
No. 182	High T_c Superconducting Thin Films, Devices, and Applications (Atlanta, GA, 1988)	88-03947	0-88318-382-X
No. 183	Cosmic Abundances of Matter (Minneapolis, MN, 1988)	89-80147	0-88318-383-8
No. 184	Physics of Particle Accelerators (Ithaca, NY, 1988)	89-83575	0-88318-384-6
No. 185	Glueballs, Hybrids, and Exotic Hadrons (Upton, NY, 1988)	89-83513	0-88318-385-4
No. 186	High-Energy Radiation Background in Space (Sanibel Island, FL, 1987)	89-83833	0-88318-386-2
No. 187	High-Energy Spin Physics (Minneapolis, MN, 1988)	89-83948	0-88318-387-0
No. 188	International Symposium on Electron Beam Ion Sources and their Applications (Upton, NY, 1988)	89-84343	0-88318-388-9
No. 189	Relativistic, Quantum Electrodynamic, and Weak Interaction Effects in Atoms (Santa Barbara, CA, 1988)	89-84431	0-88318-389-7
No. 190	Radio-frequency Power in Plasmas (Irvine, CA, 1989)	89-45805	0-88318-397-8
No. 191	Advances in Laser Science—IV (Atlanta, GA, 1988)	89-85595	0-88318-391-9
No. 192	Vacuum Mechatronics (First International Workshop) (Santa Barbara, CA, 1989)	89-45905	0-88318-394-3
No. 193	Advanced Accelerator Concepts (Lake Arrowhead, CA, 1989)	89-45914	0-88318-393-5
No. 194	Quantum Fluids and Solids—1989 (Gainesville, FL, 1989)	89-81079	0-88318-395-1
No. 195	Dense Z-Pinches (Laguna Beach, CA, 1989)	89-46212	0-88318-396-X
No. 196	Heavy Quark Physics (Ithaca, NY, 1989)	89-81583	0-88318-644-6

No. 197	Drops and Bubbles (Monterey, CA, 1988)	89-46360	0-88318-392-7
No. 198	Astrophysics in Antarctica (Newark, DE, 1989)	89-46421	0-88318-398-6
No. 199	Surface Conditioning of Vacuum Systems (Los Angeles, CA, 1989)	89-82542	0-88318-756-6
No. 200	High T_c Superconducting Thin Films: Processing, Characterization, and Applications (Boston, MA, 1989)	90-80006	0-88318-759-0
No. 201	QED Structure Functions (Ann Arbor, MI, 1989)	90-80229	0-88318-671-3
No. 202	NASA Workshop on Physics From a Lunar Base (Stanford, CA, 1989)	90-55073	0-88318-646-2
No. 203	Particle Astrophysics: The NASA Cosmic Ray Program for the 1990s and Beyond (Greenbelt, MD, 1989)	90-55077	0-88318-763-9
No. 204	Aspects of Electron-Molecule Scattering and Photoionization (New Haven, CT, 1989)	90-55175	0-88318-764-7
No. 205	The Physics of Electronic and Atomic Collisions (XVI International Conference) (New York, NY, 1989)	90-53183	0-88318-390-0
No. 206	Atomic Processes in Plasmas (Gaithersburg, MD, 1989)	90-55265	0-88318-769-8
No. 207	Astrophysics from the Moon (Annapolis, MD, 1990)	90-55582	0-88318-770-1
No. 208	Current Topics in Shock Waves (Bethlehem, PA, 1989)	90-55617	0-88318-776-0
No. 209	Computing for High Luminosity and High Intensity Facilities (Santa Fe, NM, 1990)	90-55634	0-88318-786-8
No. 210	Production and Neutralization of Negative Ions and Beams (Brookhaven, NY, 1990)	90-55316	0-88318-786-8
No. 211	High-Energy Astrophysics in the 21st Century (Taos, NM, 1989)	90-55644	0-88318-803-1
No. 212	Accelerator Instrumentation (Brookhaven, NY, 1989)	90-55838	0-88318-645-4
No. 213	Frontiers in Condensed Matter Theory (New York, NY, 1989)	90-6421	0-88318-771-X 0-88318-772-8 (pbk.)
No. 214	Beam Dynamics Issues of High-Luminosity Asymmetric Collider Rings (Berkeley, CA, 1990)	90-55857	0-88318-767-1
No. 215	X-Ray and Inner-Shell Processes (Knoxville, TN, 1990)	90-84700	0-88318-790-6

No. 216	Spectral Line Shapes, Vol. 6 (Austin, TX, 1990)	90-06278	0-88318-791-4
No. 217	Space Nuclear Power Systems (Albuquerque, NM, 1991)	90-56220	0-88318-838-4
No. 218	Positron Beams for Solids and Surfaces (London, Canada, 1990)	90-56407	0-88318-842-2
No. 219	Superconductivity and Its Applications (Buffalo, NY, 1990)	91-55020	0-88318-835-X
No. 220	High Energy Gamma-Ray Astronomy (Ann Arbor, MI, 1990)	91-70876	0-88318-812-0
No. 221	Particle Production Near Threshold (Nashville, IN, 1990)	91-55134	0-88318-829-5
No. 222	After the First Three Minutes (College Park, MD, 1990)	91-55214	0-88318-828-7
No. 223	Polarized Collider Workshop (University Park, PA, 1990)	91-71303	0-88318-826-0
No. 224	LAMPF Workshop on (π, K) Physics (Los Alamos, NM, 1990)	91-71304	0-88318-825-2
No. 225	Half Collision Resonance Phenomena in Molecules (Caracas, Venezuela, 1990)	91-55210	0-88318-840-6
No. 226	The Living Cell in Four Dimensions (Gif sur Yvette, France, 1990)	91-55209	0-88318-794-9
No. 227	Advanced Processing and Characterization Technologies (Clearwater, FL, 1991)	91-55194	0-88318-910-0
No. 228	Anomalous Nuclear Effects in Deuterium/Solid Systems (Provo, UT, 1990)	91-55245	0-88318-833-3
No. 229	Accelerator Instrumentation (Batavia, IL, 1990)	91-55347	0-88318-832-1
No. 230	Nonlinear Dynamics and Particle Acceleration (Tsukuba, Japan, 1990)	91-55348	0-88318-824-4
No. 231	Boron-Rich Solids (Albuquerque, NM, 1990)	91-53024	0-88318-793-4
No. 232	Gamma-Ray Line Astrophysics (Paris-Saclay, France, 1990)	91-55492	0-88318-875-9
No. 233	Atomic Physics 12 (Ann Arbor, MI, 1990)	91-55595	088318-811-2
No. 234	Amorphous Silicon Materials and Solar Cells (Denver, CO, 1991)	91-55575	088318-831-7

No. 235	Physics and Chemistry of MCT and Novel IR Detector Materials (San Francisco, CA, 1990)	91-55493	0-88318-931-3
No. 236	Vacuum Design of Synchrotron Light Sources (Argonne, IL, 1990)	91-55527	0-88318-873-2
No. 237	Kent M. Terwilliger Memorial Symposium (Ann Arbor, MI, 1989)	91-55576	0-88318-788-4
No. 238	Capture Gamma-Ray Spectroscopy (Pacific Grove, CA, 1990)	91-57923	0-88318-830-9
No. 239	Advances in Biomolecular Simulations (Obernai, France, 1991)	91-58106	0-88318-940-2
No. 240	Joint Soviet-American Workshop on the Physics of Semiconductor Lasers (Leningrad, USSR, 1991)	91-58537	0-88318-936-4
No. 241	Scanned Probe Microscopy (Santa Barbara, CA, 1991)	91-76758	0-88318-816-3
No. 242	Strong, Weak, and Electromagnetic Interactions in Nuclei, Atoms, and Astrophysics: A Workshop in Honor of Stewart D. Bloom's Retirement (Livermore, CA, 1991)	91-76876	0-88318-943-7
No. 243	Intersections Between Particle and Nuclear Physics (Tucson, AZ, 1991)	91-77580	0-88318-950-X
No. 244	Radio Frequency Power in Plasmas (Charleston, SC, 1991)	91-77853	0-88318-937-2
No. 245	Basic Space Science (Bangalore, India, 1991)	91-78379	0-88318-951-8
No. 246	Space Nuclear Power Systems (Albuquerque, NM, 1992)	91-58793	1-56396-027-3 1-56396-026-5 (pbk.)
No. 247	Global Warming: Physics and Facts (Washington, DC, 1991)	91-78423	0-88318-932-1
No. 248	Computer-Aided Statistical Physics (Taipei, Taiwan, 1991)	91-78378	0-88318-942-9
No. 249	The Physics of Particle Accelerators (Upton, NY, 1989, 1990)	92-52843	0-88318-789-2
No. 250	Towards a Unified Picture of Nuclear Dynamics (Nikko, Japan, 1991)	92-70143	0-88318-951-8
No. 251	Superconductivity and its Applications (Buffalo, NY, 1991)	92-52726	1-56396-016-8

No. 252	Accelerator Instrumentation (Newport News, VA, 1991)	92-70356	0-88318-934-8
No. 253	High-Brightness Beams for Advanced Accelerator Applications (College Park, MD, 1991)	92-52705	0-88318-947-X
No. 254	Testing the AGN Paradigm (College Park, MD, 1991)	92-52780	1-56396-009-5
No. 255	Advanced Beam Dynamics Workshop on Effects of Errors in Accelerators, Their Diagnosis and Corrections (Corpus Christi, TX, 1991)	92-52842	1-56396-006-0
No. 256	Slow Dynamics in Condensed Matter (Fukuoka, Japan, 1991)	92-53120	0-88318-938-0
No. 257	Atomic Processes in Plasmas (Portland, ME, 1991)	91-08105	0-88318-939-9
No. 258	Synchrotron Radiation and Dynamic Phenomena (Grenoble, France, 1991)	92-53790	1-56396-008-7
No. 259	Future Directions in Nuclear Physics with 4π Gamma Detection Systems of the New Generation (Strasbourg, France, 1991)	92-53222	0-88318-952-6
No. 260	Computational Quantum Physics (Nashville, TN, 1991)	92-71777	0-88318-933-X
No. 261	Rare and Exclusive B&K Decays and Novel Flavor Factories (Santa Monica, CA, 1991)	92-71873	1-56396-055-9
No. 262	Molecular Electronics—Science and Technology (St. Thomas, Virgin Islands, 1991)	92-72210	1-56396-041-9
No. 263	Stress-Induced Phenomena in Metallization: First International Workshop (Ithaca, NY, 1991)	92-72292	1-56396-082-6
No. 264	Particle Acceleration in Cosmic Plasmas (Newark, DE, 1991)	92-73316	0-88318-948-8
No. 265	Gamma-Ray Bursts (Huntsville, AL, 1991)	92-73456	1-56396-018-4
No. 266	Group Theory in Physics (Cocoyoc, Morelos, Mexico, 1991)	92-73457	1-56396-101-6
No. 267	Electromechanical Coupling of the Solar Atmosphere (Capri, Italy, 1991)	92-82717	1-56396-110-5
No. 268	Photovoltaic Advanced Research & Development Project (Denver, CO, 1992)	92-74159	1-56396-056-7

No. 269	CEBAF 1992 Summer Workshop (Newport News, VA, 1992)	92-75403	1-56396-067-2
No. 270	Time Reversal—The Arthur Rich Memorial Symposium (Ann Arbor, MI, 1991)	92-83852	1-56396-105-9
No. 271	Tenth Symposium Space Nuclear Power and Propulsion (Vols. I–III) (Albuquerque, NM, 1993)	92-75162	1-56396-137-7 (set)
No. 272	Proceedings of the XXVI International Conference on High Energy Physics (Vols. I and II) (Dallas, TX, 1992)	93-70412	1-56396-127-X (set)
No. 273	Superconductivity and Its Applications (Buffalo, NY, 1992)	93-70502	1-56396-189-X
No. 274	VIth International Conference on the Physics of Highly Charged Ions (Manhattan, KS, 1992)	93-70577	1-56396-102-4
No. 275	Atomic Physics 13 (Munich, Germany, 1992)	93-70826	1-56396-057-5
No. 276	Very High Energy Cosmic-Ray Interactions: VIIth International Symposium (Ann Arbor, MI, 1992)	93-71342	1-56396-038-9
No. 277	The World at Risk: Natural Hazards and Climate Change (Cambridge, MA, 1992)	93-71333	1-56396-066-4
No. 278	Back to the Galaxy (College Park, MD, 1992)	93-71543	1-56396-227-6
No. 279	Advanced Accelerator Concepts (Port Jefferson, NY, 1992)	93-71773	1-56396-191-1
No. 280	Compton Gamma-Ray Observatory (St. Louis, MO, 1992)	93-71830	1-56396-104-0
No. 281	Accelerator Instrumentation Fourth Annual Workshop (Berkeley, CA, 1992)	93-072110	1-56396-190-3
No. 282	Quantum 1/f Noise & Other Low Frequency Fluctuations in Electronic Devices (St. Louis, MO, 1992)	93-072366	1-56396-252-7
No. 283	Earth and Space Science Information Systems (Pasadena, CA, 1992)	93-072360	1-56396-094-X

No. 284	US-Japan Workshop on Ion Temperature Gradient-Driven Turbulent Transport (Austin, TX, 1993)	93-72460	1-56396-221-7
No. 285	Noise in Physical Systems and 1/f Fluctuations (St. Louis, MO, 1993)	93-72575	1-56396-270-5
No. 286	Ordering Disorder: Prospect and Retrospect in Condensed Matter Physics: Proceedings of the Indo-U.S. Workshop (Hyderabad, India, 1993)	93-072549	1-56396-255-1
No. 287	Production and Neutralization of Negative Ions and Beams: Sixth International Symposium (Upton, NY, 1992)	93-72821	1-56396-103-2
No. 288	Laser Ablation: Mechanismas and Applications-II: Second International Conference (Knoxville, TN, 1993)	93-73040	1-56396-226-8
No. 289	Radio Frequency Power in Plasmas: Tenth Topical Conference (Boston, MA, 1993)	93-72964	1-56396-264-0
No. 290	Laser Spectroscopy: XIth International Conference (Hot Springs, VA, 1993)	93-73050	1-56396-262-4
No. 291	Prairie View Summer Science Academy (Prairie View, TX, 1992)	93-73081	1-56396-133-4
No. 292	Stability of Particle Motion in Storage Rings (Upton, NY, 1992)	93-73534	1-56396-225-X
No. 293	Polarized Ion Sources and Polarized Gas Targets (Madison, WI, 1993)	93-74102	1-56396-220-9
No. 294	High-Energy Solar Phenomena A New Era of Spacecraft Measurements (Waterville Valley, NH, 1993)	93-74147	1-56396-291-8
No. 295	The Physics of Electronic and Atomic Collisions: XVIII International Conference (Aarhus, Denmark, 1993)	93-74103	1-56396-290-X
No. 296	The Chaos Paradigm: Developments an Applications in Engineering and Science (Mystic, CT, 1993)	93-74146	1-56396-254-3
No. 297	Computational Accelerator Physics (Los Alamos, NM, 1993)	93-74205	1-56396-222-5